T0324142

Springer Theses

Recognizing Outstanding Ph.D. Research

Aims and Scope

The series "Springer Theses" brings together a selection of the very best Ph.D. theses from around the world and across the physical sciences. Nominated and endorsed by two recognized specialists, each published volume has been selected for its scientific excellence and the high impact of its contents for the pertinent field of research. For greater accessibility to non-specialists, the published versions include an extended introduction, as well as a foreword by the student's supervisor explaining the special relevance of the work for the field. As a whole, the series will provide a valuable resource both for newcomers to the research fields described, and for other scientists seeking detailed background information on special questions. Finally, it provides an accredited documentation of the valuable contributions made by today's younger generation of scientists.

Theses are accepted into the series by invited nomination only and must fulfill all of the following criteria

- They must be written in good English.
- The topic should fall within the confines of Chemistry, Physics, Earth Sciences, Engineering and related interdisciplinary fields such as Materials, Nanoscience, Chemical Engineering, Complex Systems and Biophysics.
- The work reported in the thesis must represent a significant scientific advance.
- If the thesis includes previously published material, permission to reproduce this must be gained from the respective copyright holder.
- They must have been examined and passed during the 12 months prior to nomination.
- Each thesis should include a foreword by the supervisor outlining the significance of its content.
- The theses should have a clearly defined structure including an introduction accessible to scientists not expert in that particular field.

More information about this series at http://www.springer.com/series/8790

Cheng-Meng Chen

Surface Chemistry and Macroscopic Assembly of Graphene for Application in Energy Storage

Doctoral Thesis accepted by
University of Chinese Academy of Sciences, China

Author
Dr. Cheng-Meng Chen
Institute of Coal Chemistry
Chinese Academy of Sciences
Taiyuan
People's Republic of China

Supervisors
Prof. Yong-Gang Yang
Institute of Coal Chemistry
Chinese Academy of Sciences
Taiyuan
People's Republic of China

Prof. Quan-Hong Yang
School of Chemical Engineering and Technology
Tianjin University
Tianjin
People's Republic of China

ISSN 2190-5053 ISSN 2190-5061 (electronic)
Springer Theses
ISBN 978-3-662-48674-0 ISBN 978-3-662-48676-4 (eBook)
DOI 10.1007/978-3-662-48676-4

Library of Congress Control Number: 2015953237

Springer Heidelberg New York Dordrecht London

Printed on acid-free paper

Springer-Verlag GmbH Berlin Heidelberg is part of Springer Science+Business Media
(www.springer.com)

Parts of this thesis have been published in the following journal articles:

1. Cheng-Meng Chen; Qiang Zhang; Mang-Guo Yang; Chun-Hsien Huang; Yong-Gang Yang; Mang-Zhang Wang; Structural evolution during annealing of thermally reduced graphene nanosheets for application in supercapacitors. *Carbon* **2012**, *50*, 3572. Chapter 2, Copyright 2012, Reprinted with permission from Elsevier.
2. Cheng-Meng Chen; Qiang Zhang; Xiao-Chen Zhao; Bing-Sen Zhang; Qing-Qiang Kong; Mang-Guo Yang; Quan-Hong Yang; Mao-Zhang Wang; Yong-Gang Yang; Robert Schlögl; Dang-Sheng Su. Hierarchically aminated graphene honeycombs for electrochemical capacitive energy storage. *J. Mater. Chem.* **2012**, *22*, 14076. Chapter 3, Reprinted by permission of The Royal Society of Chemistry.
3. Cheng-Meng Chen; Jia-Qi Huang; Qiang Zhang; Wen-Zhan Gong; Quan-Hong Yang; Mao-Zhang Wang; Yong-Gang Yang. Annealing a graphene oxide film to produce a free standing high conductive graphene film. *Carbon* **2012**, *50*, 659. Chapter 4, Copyright 2012, Reprinted with permission from Elsevier.
4. Cheng-Meng Chen; Qiang Zhang; Chun-Hsien Huang; Xiao-Chen Zhao; Bing-Sen Zhang; Qing-Qiang Kong; Mao-Zhang Wang; Yong-Gang Yang; Rong Cai; Dang-Sheng Su. Macroporous 'bubble' graphene film via template-directed ordered-assembly for high rate supercapacitors. *Chem. Commun.* **2012**, *48*, 7149. Chapter 5, Reprinted by permission of The Royal Society of Chemistry.
5. Cheng-Meng Chen; Qiang Zhang; Jia-Qi Huang, Wei Zhang, Xiao-Chen Zhao; Chun-Hsien Huang; Fei Wei; Yong-Gang Yang; Mao-Zhang Wang; Dang-Sheng Su. Chemically derived graphene–metal oxide hybrids as electrodes for electrochemical energy storage: pre-graphenization or post-graphenization? *J. Mater. Chem.* **2012**, *22*, 13947. Chapter 6, Reprinted by permission of The Royal Society of Chemistry.

Supervisors' Foreword

Graphene is a wonder material with many superlative performances. As a rapidly rising star, graphene has attracted enormous efforts and vast investments since the reception of Nobel Prize in 2010. It offers the promise to significantly improve the performances of existing products and to enable the design of materials and devices with novel functionalities. Notable progresses have been made on the preparation, property, and application exploration of graphene. However, there are still challenges remained for the research, i.e., surface chemistry, structure modification, macroscopic assembly, and practical application.

Aiming at the above critical subjects, Dr. Cheng-Meng Chen's doctoral thesis describes the tunable surface chemistry of graphene/graphene oxide and their macroscopic assemblies/hybrids in the applications of advanced energy storage and conversion. The surface functional groups of thermally reduced graphene nanosheets prepared by vacuum-promoted thermal expansion of graphene oxide are tailored by progressive carbonization. The hierarchically aminated graphitic honeycombs (AGHs) are fabricated for electrochemical capacitive energy storage application through a facile high-vacuum-promoted thermal expansion and subsequent amination process. The structural evolution of graphene, correlated with electrochemical performance, is explored based on tunable surface chemistry.

Dr. Chen's research also bridges between the surface chemistry and the electrochemical property of graphene and laid the theoretical foundation for developing graphene-based energy storage devices. Working as precursor, graphene oxide is self- or template-assembled into macrostructures followed by annealing at varied temperatures. A free-standing graphene oxide film (GOF) obtained by self-assembly at a liquid/air interface is annealed in a confined space between two stacked substrates to form a free-standing highly conductive graphene film. An ultralight 3D macroporous bubble graphene film is obtained by a facile hard template-directed ordered assembly approach. Secondary phase (SnO_2) is introduced into graphene system to improve the electrochemical performance. The differences in the structure and electrochemical properties of pre- and post-graphenized graphene/SnO_2 hybrids are investigated. Dr. Chen's work provides a general strategy for graphene macrostructure toward other applications in biology, energy conversion, composite, and catalysis, which requires tunable pore and tailorable microstructures.

It is supposed in this doctoral thesis that researches on surface chemistry and macroscopic assembly open wide space for the chemically derived graphene toward the applications in advanced energy conversion and storage. There are lots of work can be done to further understand the graphene-related chemistry and application.

Taiyuan
Tianjin
May 2015

Prof. Yong-Gang Yang
Prof. Quan-Hong Yang

Acknowledgments

This dissertation is completed under the guidance of Professor Yong-Gang Yang in Institute of Coal Chemistry, Chinese Academy of Sciences and Quan-Hong Yang in Tianjin University. I would like to express my deepest gratitude to my advisors for their patience and guidance in research, as well as meticulous care and attention in life. This work, from topic selection to experiment and dissertation preparation, cannot go well without their devotion. Professor Yong-Gang Yang makes great contribution to our country, and his dedication to carbon fiber and attitude toward research encourage and boost my motivation. His care and tolerance affect me a lot. Professor Quan-Hong Yang has deep understanding in nanocarbon materials (graphene, etc.). His enthusiasm in research and insightful views in carbon materials benefit me a lot.

I am thankful for the support and encouragements of Professor Mao-Zhang Wang. He keeps working in scientific research in his seventies. His enthusiasm encourages me to fight my way. His wide knowledge and vision, rigorous attitude, and modest affability help me a lot to confront the future life with open and tolerate mind.

I would like to express my sincere thanks to Professor Dang-Sheng Su (Former Supervisor of Electron Microscope Laboratory in Department of Inorganic, Fritz Haber Institute, Max Planck Society) for providing opportunity to go abroad for further study. During study in German, Professor Su provided me with not only the first-class experimental condition and research atmosphere, but also the invaluable scientific attainments and study methods. It has been an invaluable experience to work under his edification. In addition, I am thankful for the guidance and help of Professor Robert Schlögl, Department Chairman, for his result discussion and manuscript revision.

I am grateful to meet and cooperate with the following young talents. My thanks go to Professor Qiang Zhang from Tsinghua University for his guidance in research and encouragement in life. It is my great honor to meet him. My thanks go to Dr. Xiao-Chen Zhao, from Dalian Institute of Chemical Physics, Chinese Academy of Sciences, for her help in energy storage and conversion. My thanks go to

Professor Bing-Sen Zhang, from Institute of Metal Research, Chinese Academy of Sciences, for his help in electron microscope characterization and daily life. My thanks also go to Jun-Xian Huang, from National Tsinghua University, for his discussion in research idea and result discussion. Moreover, I would like to thank my colleagues and friends in Fritz Haber Institute, Wei-Qing Zheng, Wei Zhang, Lian-Di Li, He-Nan Li, Sylvia, Youngmi, Till, Adriana, Klaus, Julian and Bärbel, for their support and help.

I am thankful for the support and encouragements of Group 701, especially for the help of Qing-Qiang Kong and Mang-Guo Yang in preparing graphene samples, Xing-Hua Zhang in research idea, Yang Li and Wen-Zhao Gong in life, as well as other colleagues, Shou-Chun Zhang, Hui Li, Jun-Liang Fan, Pei-Xian Hu, Gong-Qiu Peng, Jian-Guo Gao, Fang-Yuan Su, Wei Wang, Tiao-Lan Zhang, Ping Wang, Li-Na Wang, Gang-Ling Xu, Run-E Wang, Shi-Yao Feng, Yi Jiang, Kai Chen and Wei-Ning Zhang et al.

I am grateful to Professor Rong Cai and Jian-Guo Wang et al. for their great support for graphene-related research work. I am thankful for the support and help from Key Laboratory of Carbon Materials, Graduate School, Science and Technology Department, Material Department, Analysis and Test Center, Laboratory, Editorial Department of <New Carbon Materials>, etc. I am also thankful for the guidance of cooperators, Professor Feng Li and Hui-Ming Cheng in Institute of Metal Research, Chinese Academy of Sciences.

My special thanks go to my parents and sister for their support and encouragement. It's your endless love and dedication that give me the impetus to move forward.

Last but foremost, I would like to acknowledge Ms. Zhuo Liu for her helpful editorial and translation work on this thesis and express many thanks to all the scholars and professionals who paid hard in reviewing this dissertation.

Contents

Chapter 1
Literature Review and Research Background

1.1 Introduction to Graphene

Graphene, a two-dimensional crystal of sp^2 hybridized carbon atoms which are one-atom thick, is the thinnest artificial material at present. It is the basic building block for fullerenes (0D), carbon nanotube (1D), and graphite (3D) (Fig. 1.1) [1]. The research on monolayer graphite by chemists and surface scientists can be traced back to 1860s. Boehm et al. named this monolayer graphite as graphene in 1986 [2]. However, the two-dimensional crystal structure was thought to be thermodynamic instability at its free state, which cannot exist independently in general environment [3]. Until 2004, Andre Geim and Konstantin Novoselov, physicists at University of Manchester, peeled off small amount of monolayer graphene by adhesive tape, obtaining free-standing graphene in air, and changed the quantum relativity theory of 'concept' material into reality [4]. They shared the Nobel Prize in Physics 2010 for their groundbreaking work in 2D graphene material [5].

Graphene is not only the thinnest material ever tested, but also the strongest nanomaterial (tensile strength $E \approx 1.01$ TPa, ultimate strength $\sigma \approx 130$ GPa) [6]. It is almost completely transparent, absorbing only 2.3 % light [7]. Its theoretical specific surface area is as high as 2630 $m^2\ g^{-1}$, thermal conductivity as high as 5300 W $m^{-1}\ K^{-1}$, which is higher than that of carbon nanotube and diamond, the electron mobility as high as 15,000 $cm^2\ V^{-1}\ S^{-1}$ at room temperature, which is higher than that of carbon nanotubes or silicon crystal. However, the electrical resistivity of graphene is only about $10^{-6}\ \Omega$ cm, lower than that of copper or silver, being the lowest resistivity material in the world. Another character of graphene is the quantum Hall effect observed at room temperature (Fig. 1.1). Graphene is expected to work in developing new electronic devices and transistors with low thickness and high conductive speed, based on its very low electrical resistivity and high traveling speed of electrons. It was even expected to be new semiconducting material to replace silicon [8]. Graphene is substantially a good transparent conductor, which is suitable for making transparent touch screen, light guide plate, and solar battery [9].

C.-M. Chen, *Surface Chemistry and Macroscopic Assembly of Graphene for Application in Energy Storage*, Springer Theses,
DOI 10.1007/978-3-662-48676-4_1

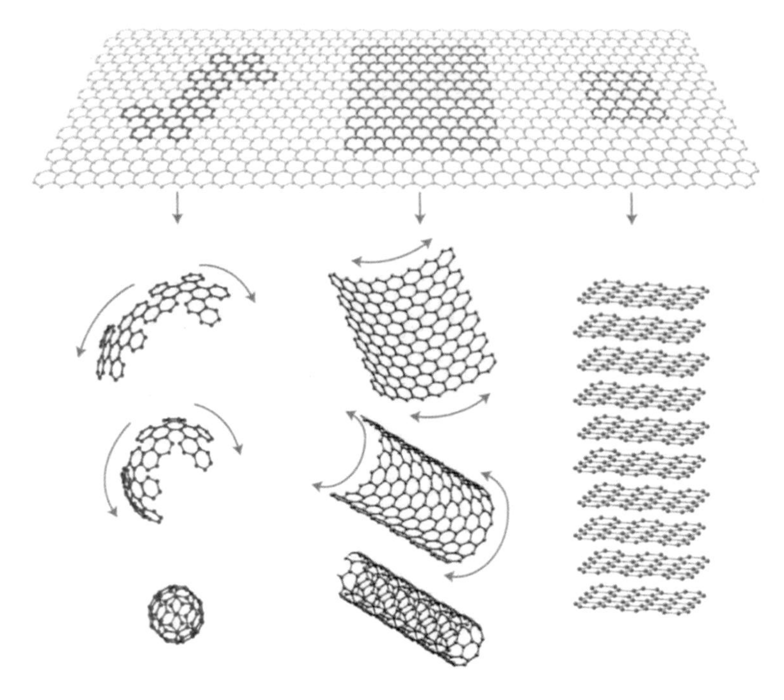

Fig. 1.1 Graphene—the basic building block of graphitic carbon. Reprinted by permission from Macmillan Publishers Ltd.: Ref. [1], copyright 2007

1.2 Summary of Graphene Preparation Methods

Controllable and scalable preparation is the precondition and foundation of commercial application of graphene. The preparation of free-state graphene can be divided into physical and chemical methods. In the aspect of chemical nanomaterial, it can be divided into dissociating graphite from top to bottom and chemical synthesis from bottom to top (Fig. 1.2). The present researched methods can be roughly classified as follows: (1) micromechanical exfoliation, (2) epitaxial growth, (3) chemical vapor deposition, (4) vertical split of carbon nanotubes, (5) reduction of graphene derivatives, (6) chemical synthesis, etc. In fact, the current methods are still unable to realize mass production of defect-free monolayer graphene.

As for preparing graphene powders in gram level, or single-layer graphene with area ≥ 2 cm^2 to fabricate devices, methods 1–4 cannot satisfy the requirements. Adhesive tape stripping method is hard to operate, and it is less likely to obtain graphene sheet with good performance. The epitaxial growth method helps to obtain graphene in high quality, but requires high vacuum and special equipment. It is used to prepare film in small area only. The development of chemical vapor deposition paves way for producing single-layer graphene in large area, but it still needs further development. Similarly, although vertical split of carbon nanotubes can be used to

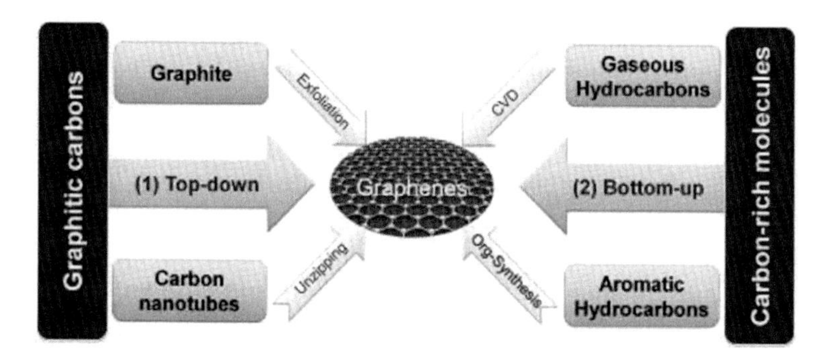

Fig. 1.2 Chemical approaches toward preparation of graphene. Reproduced from Ref. [10] by permission of John Wiley & Sons Ltd.

prepare graphene stripe in large scale, the width of stripe depends on the diameter of carbon tube. This method is newly explored, so the future of large-scale production remains to be proved.

Graphite is natural mineral formed in harsh geological mineralization condition. Essentially, it is stacked by numerous graphene sheets, expected to be the precursor material for mass production of graphene. Graphene is composed of six-membered rings through sp^2 hybridization in two-dimensional direction, and in bulk graphite the neighboring graphene layers arrange in overlapped p_z orbital, with strong van der Waals force (300 nN μm^{-2}) between layers [11]. It is hard to completely exfoliate graphite into single-layer graphene, generally obtained the few-layer stacked sheets and tiny amount of single-layer sheets. Chemical method works through chemical intercalation of graphite oxide in liquid, obtaining graphite oxide. In this condition, the d_{002} interlayer space of graphite oxide is increased, corresponding to decreased interlayer van der Waals force. Graphene oxide (GO) hydrosol or functional graphene single layer can be easily obtained by ultrasonic treatment or thermal expanded exfoliation. Further chemical reduction or annealing can realize the conversion to graphene (Fig. 1.3). This method has the advantages of easily available raw materials, low cost, technical maturity, etc., which are considered to be the starting point of wide application of graphene.

Falling between large molecule and nanoscale, chemically obtained graphene (strictly speaking, 'highly reduced graphene oxide') contains amount of functional groups and defects, which restricts its application in quantum physics and electronic field. However, this kind of graphene can be modified through chemical reaction to be dispersed in various medium, forming organic or inorganic nanocomposites based on its controllable surface chemistry. At the same time, high workability in wet chemical condition allows it to be the nanounit for assembling paper-, film-, or foam-like materials, showing application potential in energy storage/conversion, catalysis, etc. Therefore, the following will comprehensively review the preparation, structure,

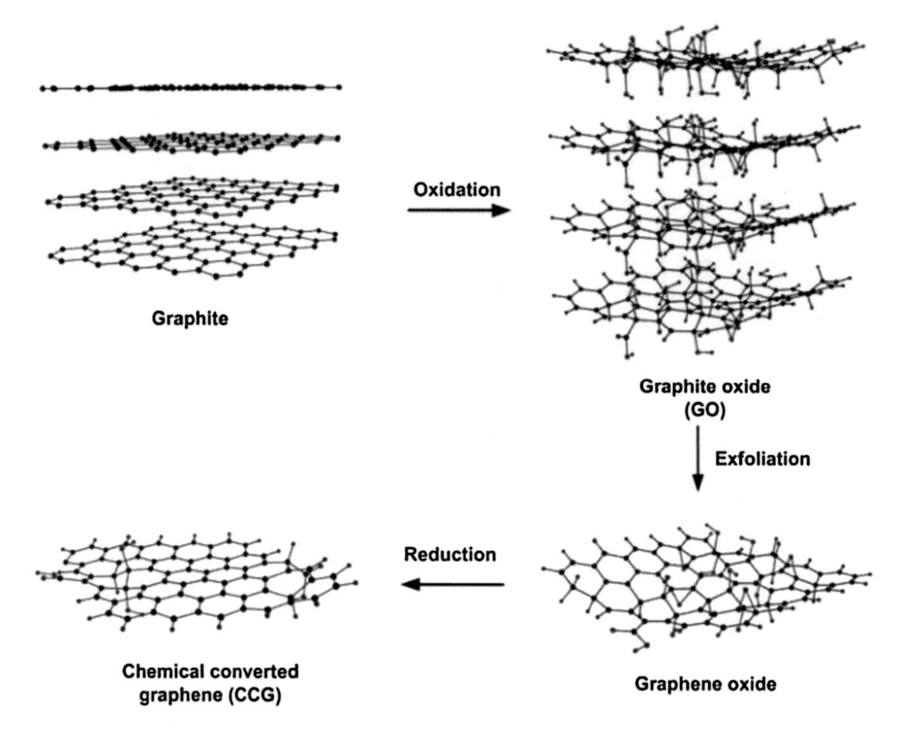

Fig. 1.3 Preparation of chemically converted graphene (CCG) by reduction of graphene oxide. Reproduced from Ref. [12] by permission of John Wiley & Sons Ltd.

and reactivity of graphene (including deoxygenation and chemical modification), in the aspect of surface chemistry of graphene oxide.

1.3 The Synthesis and Structure of Graphite Oxide

As potential new material, graphene has its novelty, but graphite oxide has a long history (that is, graphite chemistry). In 1859, Brodie, a British chemist, explored the structure of graphite by studying the reactivity of flake graphite. Brodie clamed to obtain the material composed of carbon, hydrogen, and oxygen by adding $KClO_3$ into graphite suspension in nitrosonitric acid, and the total mass of resultants increased. But he found that the lattice angle of the above-separated crystal cannot be measured by XRD. Continuous oxidation treatment resulted in further increased oxygen content, which reached the limit after four times reaction with molecular formula $C_{2.19}H_{0.80}O_{1.00}$. This material can be dispersed in neutral or alkaline aqueous solution, but not the acid medium, so he named it graphite acid. The composition of this material changed into $C_{5.51}H_{0.48}O_{1.00}$ after heating to 220 °C, accompanied by

Table 1.1 The methods to oxidize graphite into graphene

	Brodie	Staudenmaier	Hummers	Modified Hummers	
Years	1859	1898	1958	1999	2004
Oxidant	$KClO_3$, HNO_3	$KClO_3$ (or $NaClO_3$), HNO_3, H_2SO_4	$NaNO_3$, $KMnO_4$, H_2SO_4	Preoxidation: $K_2S_2O_8$, P_2O_5, H_2SO_4 oxidation: $KMnO_4$, H_2SO_4	$NaNO_3$, $KMnO_4$, H_2SO_4
C:O (Atomic ratio)	2.16	N/A	2.25	1.3	1.8
	2.28	1.85	2.17		
Reaction time	3–4 days	1–2 days	About 2 h	Preoxidation 6 h + Oxidation 2 h	About 5 days
	10 h	10 days	9–10 h		
Layer spacing (Å)	5.95	6.23	6.67	6.9	8.3

Reproduced from Ref. [13] by permission of John Wiley & Sons Ltd.

the loss of H_2O, CO_2, and CO. Brodie was always interested in the molecular formula and weight of 'graphite', and finally decided the molecular weight as 33. He supposed that 'this carbon requires a new name to reveal it is a new element, and I suggest naming it Graphon'.

At present, nearly 150 years later, graphene storm has swept the entire physical and chemical fields, and now we know that the molecular formula predicted by Brodie is wrong, as the uncertainty of this material is beyond his imagination. 40 years later, after the discovery of graphite oxide by Brodie, L. Staudenmaier slightly changed the process by adding equal amount of chlorate for several times, not once as Brodie did. But the oxidation degree was comparable with that oxidized several times by Brodie. The real advantage is it realizes industrialized preparation by single reactor.

60 years later after Staudenmaier's method, Hummers and Offeman developed a new method. This method reacted graphite with potassium permanganate and concentrated sulfuric acid, getting similar oxidation result of the former method. On this basis, improved methods were proposed subsequently, but the synthesis routes of graphite oxide were all limited to the above three. However, the corresponding performances were different from each other. These differences not only related with the used oxidant, but also depended on the used raw material and reaction condition. The reaction conditions are summarized in Table 1.1. At present, we know little about the direct reaction mechanism of each method, but can be inspired by the general reactivity of involved chemical agents in the reaction. Brodie's and Staudenmaier's methods used $KClO_3$ and nitric acid, so these two methods were discussed together.

Nitric acid is a common oxidizing agent, and it can react with various aromatic carbon (including carbon nanotube, etc.). The reaction forms different oxygen-containing species, such as carboxyl, lactone, ketone, etc. Gaseous NO_2 and N_2O_4, i.e., the yellow gas described by Brodie, are released during oxidation by nitric acid. Potassium perchlorate is a strong oxidant, commonly used in detonator or other explosive materials. The twin oxidation sites are major reaction species. These are the strongest oxidation conditions at that time, as well as one of the severest reaction conditions in present laboratory scale.

Hummer's method combined potassium permanganate and sulfuric acid. Although potassium permanganate is a common oxidizing agents, the actual reactant was Mn_2O_7 (Eqs. 1.1 and 1.2), a dark red oily-like material formed by reaction between $KMnO_4$ and H_2SO_4 [14]. The reactivity of double metal heptoxide is much higher than that of monometal tetroxide, and the former may explode when heating at 55 °C (or higher) or contacts with organic compounds. Tromel and Russ pointed out that, relative to aromatic double bond, Mn_2O_7 can selectively oxidize the unsaturated aliphatic double bond. These provide important clues for studying the structure and oxidation reaction route of graphite.

$$KMnO_4 + 3H_2SO_4 \rightarrow K^+ + MnO_3^+ + H_3O^+ + 3HSO_4^- \quad (1.1)$$

$$MnO_3^+ + MnO_4^+ \rightarrow Mn_2O_7 \quad (1.2)$$

Flake graphite is the commonly used source material in chemical reaction (including graphite oxidation). It is a kind of naturally occurring mineral, which needs purification to remove undesired atoms. Similarly, there are a lot of defects in its π structure, and these defects can work as starting point for oxidation. Learning from Tromel and Russ's research on styrene, it is possible that the oxidation occurs in the isolated alkene, not the aromatic system. However, the complexity of flake graphite and its inherent defects make it difficult to precisely explain the oxidation mechanism. In addition, almost no other oxidants are used in the synthesis of graphite oxide. Jones reagent (H_2CrO_4/H_2SO_4) is commonly used in the synthesis of expanded graphite, but its partial oxidation and intercalation structure actually is between graphite and true graphite oxide. Wissler reviewed the sources of various graphite and carbon materials, and described the terminologies used in these materials, which are of great reference value [15].

The sp^2 hybridized structure in stacked graphite sheet is destroyed when the graphite is oxidized into GO, forming wrinkled defects. These defects increase the interlayer space from 3.35 Å in graphite powder to 6.8 Å in GO powder, which depends on the amount of intercalated water and the decreased interaction between sheets. So GO decomposed into single-layer graphene oxide by low-energy ultrasonic treatment in water. The surface of graphene oxide is hydrophilic and negatively charged with oxygen-containing functional groups in the neutral condition, which can be stably dispersed in water. This dispersion system is the precursor for preparing various graphene oxide and graphene-based materials.

In addition to the mechanism of effective oxidation reaction, the exact chemical structure of graphite oxide is also controversial in recent years. Up to now, there is no definite model. Many reasons may exist in this subject, but the main factor is the complexity (including the diversities among different samples) and the lack of exact analysis technology to characterize this material (or derived compound materials). The complexity of the material roots in its amorphous form and non-stoichiometric property. However, the exploration of graphite oxide structure is continued and obtains great success, in spite of the above difficulty.

The early structural model is usually ordered crystal lattice composed by discrete and repeated units. Hofmann and Holst's structure (Fig. 1.4) is composed by epoxy functional groups dispersed on the basal plane of graphite, and the simplified molecular formula is C_2O. Ruess proposed a variant of this model in 1946 by introducing hydroxyl groups into basal plane, thus explaining the H content in graphite oxide. Ruess's model changed the structure of basal plane into sp^3 hybridized system, different from Hofmann and Holst's sp^2 hybridized model. Ruess's model assumed that the sp^3 hybridization was the singly repeated unit, among which a quarter of cyclohexane was composed by 1,3-epoxy groups and 4-hydroxy, forming well-structured crystal. This assumption was confirmed by Mermoux's observation on the structural similarity of graphite oxide and polyfluorinated carbon $(CF)_n$, which referenced the forming mechanism of C–F bond and used it to explain the transfer of sp^2 hybridization in graphite into sp^3 structure of cyclohexane. Scholz and Boehm proposed a new model in 1969, which replaced the epoxy groups and ether groups completely with ordered benzoquinone [17]. Another famous model is proposed by Nakajima and Matsuo. This model assumed a crystal framework similar to poly-perfluorocarbon, and derived second-order graphite interlayer compound (GIC) based on the model [18]. These authors built progressive structural models based on three different oxidation methods of preparing GO, making great contribution in understanding the chemical performances of GO [19].

Recently, the graphite oxide model is no longer limited to simple crystal lattice model, but concentrated in amorphous structure without stoichiometric ratio. Of course, the most famous model was proposed by Lerf and Klinowski (Fig. 1.5). They published several papers about the structure of graphite oxide and its hydrate behavior, which are widely cited at present. The original work by Lerf and his colleague was to characterize this material with solid nuclear magnetic resonance spectroscopy (NMR), which was the first time in this field, as the former model was mainly obtained by element composition, reactivity, and XRD research. By preparing a serious of derivatives from graphite oxide, Lerf acquired the structural character based on the reactivity of material.

In the ^{13}C nuclear magnetic resonance spectrum of graphite oxide, there are three broad resonance peaks at 60, 70, 130 ppm displayed in the cross polarization/magic angle spinning experiment (CP/MAS), while the signal of short contact time spectra was only captured at 60 and 70 ppm. Mermoux's model showed that the chemical valence of carbon atom in graphite oxide was four, with the peak at 60 ppm corresponding to tertiary alcohols, 70 ppm to epoxy functional group, and 130 ppm to the mixture of alkene. Short contact time experiments also showed that strong interlayer

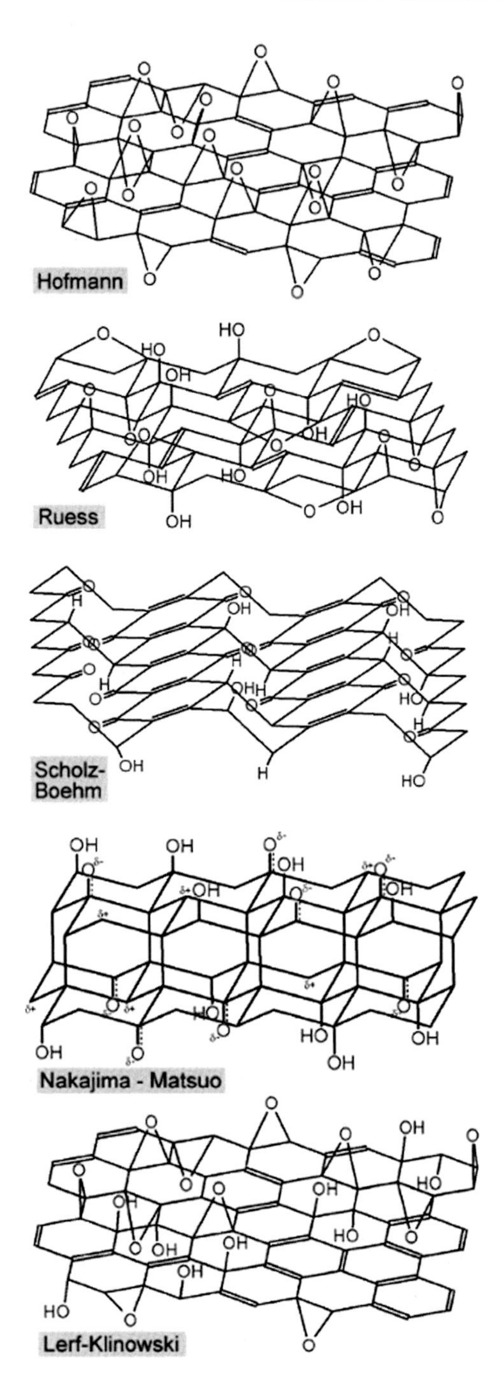

Fig. 1.4 Summary of several older structural models of GO. Reprinted with the permission from Ref. [16]. Copyright 2006 American Chemical Society

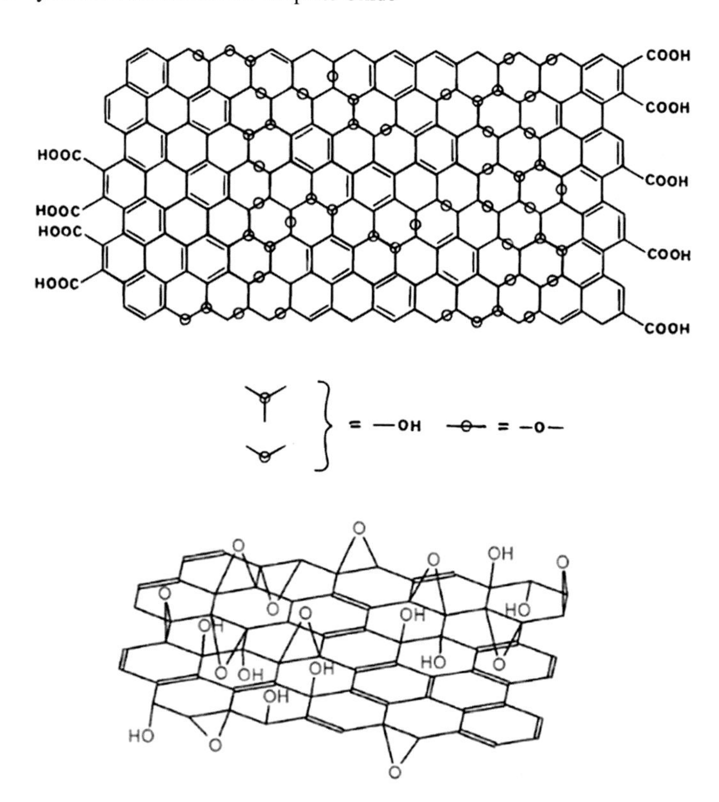

Fig. 1.5 Variations of the Lerf–Klinowski model indicating ambiguity regarding the presence (*top* reprinted with the permission from Ref. [19]. Copyright 1998 American Chemical Society.) or absence (*bottom* reprinted with the permission from Ref. [16]. Copyright 2006 American Chemical Society.) of carboxylic acids on the periphery of the basal plane of the graphitic platelets of GO

hydrogen bonds were formed between the epoxy and hydroxyl groups, forming the stacking structure of graphite oxide. These results agreed with the overall functional groups in the old models, but it was still uncertain about the dispersion of these functional groups. Especially for alkene, was it located isolated, or aggregated in conjugated or aromatic clusters? The answer to this question plays an important role in studying the electronic structure and chemical reactivity of GO.

To solve this problem, Lerf and his colleague reacted graphite oxide with cis-butenedioic anhydride [20]. Cis-butenedioic anhydride is commonly used dienophile for [4 + 2] Diels–Alder cycloaddition reaction, and the conjugated nonaromatic alkene is liable to react with this substrate. However, ^{1}H and ^{13}C spectra are almost the same with the original material, indicating that no reaction takes place. The results are not accurate to some degree.

The neutron signal may be annihilated by the strong resonance, resulting from the adsorption of water on GO surface. However, considering the resolution of the neutron signal, the interference from water peak can be eliminated from the corresponding ^{1}H NMR spectrum by treating GO with D_2O, while the signal of tertiary alcohol is unaffected. So it is a slow exchange process for water molecules inserting into GO interlayer. There is also a peak at 1.0 ppm, indicating the coexistent of at least two kinds of alcohols with equivalent magnetism. It is still uncertain about this material, but can be speculated that strong hydrogen bond exists between interlayers or interlayer-inserted water. Reaction with sodium ethoxide showed that ethylate can be the electrophilic center of surface functionalization [20, 21]. It can be seen that the major structure of GO is composed by tertiary alcohol and ethers, especially the 1,2-ether (such as epoxy group). The above conclusions are the research foundation of GO reactivity, and will be further discussed in detail.

It should be noted that, in the early NMR research, the half peak width of water was almost stable at 123–476 K, which showed that strong reaction occurred between water and GO [21] (Fig. 1.6). This may be the key factor for maintaining the stacked structure of GO. Besides, the behavior of water in graphite oxide has been characterized by neutron scattering, proving that water can be intensely connected with the GO basal plane by forming hydrogen bond with the oxygen in epoxy group [22–24].

The above researches have described the fundamental structure feature of GO, but it is necessary to construct an accurate model covering the complexity of GO. Lerf et al. reacted GO with a series of reagents [19], further confirming that the double bond existed in aromatic or conjugate structure, but not independently existed in strong oxygen condition (modified Hummers' method). Newer Lerf–Klinowski model (Fig. 1.5) imported IR spectra data from 10 years ago, indicating that in addition to ketone groups, small amount of carboxyl groups are distributed in the edge of graphene layer [17, 25, 26].

Similar to Brodie's research in 1859 and other researches later, Lerf and his colleagues noticed the thermal instability of GO. The signals of ^{13}C NMR spectrum at 60 and 70 ppm disappeared, leaving behind signal at 122 ppm, after annealing GO at 100 °C in vacuum. This signal was ascribed to aromatic and phenol compounds [28]. It is well known that the decomposition of GO releases CO and CO_2, not O_2, because of the inherently high surface reactivity of GO [29]. Oxygen-containing graphitized carbon, highly disordered compound, was obtained when GO was discomposed at high temperature. This graphitized carbon was hard to characterize. Lerf–Klinowski's model kept almost unchanged after being proposed for 10 years, but small changes on this model were made by other researchers, including the proposed existence of five-membered and six-membered lactols on the edge of graphite plane, the grafting of tertiary alcohol ether, etc. All these studies were based on the common view that the basal plane was mainly composed by epoxides and alcohols [30, 31]. Cai et al. displayed the isotope labeled GO, and broadened the application of spectrum technology, which can be used in researching GO structure [30].

Dekany and his colleagues suggested an important model different from Lerf–Klinowski's model, renewing Ruess and Scholz–Boehm's models. They updated

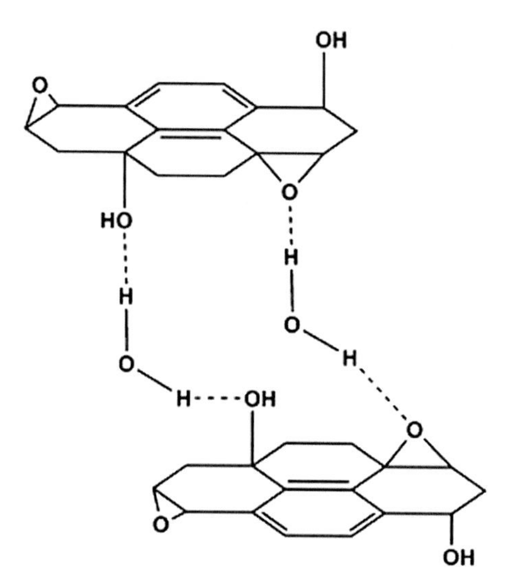

Fig. 1.6 Proposed hydrogen bonding network formed between oxygen functionality on GO and water. Reprinted with the permission from Ref. [19]. Copyright 1998 American Chemical Society. Reprinted from Ref. [21], Copyright 1997, with permission from Elsevier. Reproduced from Ref. [27] by permission of The Royal Society of Chemistry

these two models, indicating that the ordered conjugated quinone group was broken by tertiary alcohol and 1,3-ether functionalized trans-cyclohexyl. Combining the infrared signature and DRIFT spectrum of GO [32], it was found that the infrared peak at 1714 cm^{-1} was ketone group or quinone group, not carboxylic group. However, the acid–base potentiometric titration showed that acid site existed on GO plane [33]. To explain the difference, ketone–enol in situ isomerization, brought by unsaturated α, β ketone, was introduced into Lerf–Klinowski's model, and the enol sites were used to provide proton exchange sites. However, it is thermodynamically reasonable for ketone, but not reasonable for enolization and acidic proton transfer. Of course, if enol existed in aromatic area (for example, the exchange between phenol and quinone), phenoxide would be the thermodynamic product, and the proton transfer was permitted.

In conclusion, Dekany's model (Fig. 1.7) was composed of two major parts: cyclohexyl part, covered by trans-tertiary alcohol and 1,3-ether, and conjugated ketone/quinone groups; and carboxyl was not accepted in his description. Further oxidation formed 1,2-ether, and destroyed the alkene in benzoquinone, while aromatic character still existed in the initial oxidation stage. Quinones were considered to introducing rigid plane edge, which might be the source of commonly observed microwrinkle in TEM image of GO [16].

At last, the term of graphite is changeable and misleading, as different precursors (mainly the graphite source) and/or oxidation methods corresponded to different oxidation degree, further resulting in different structures and performances of GO.

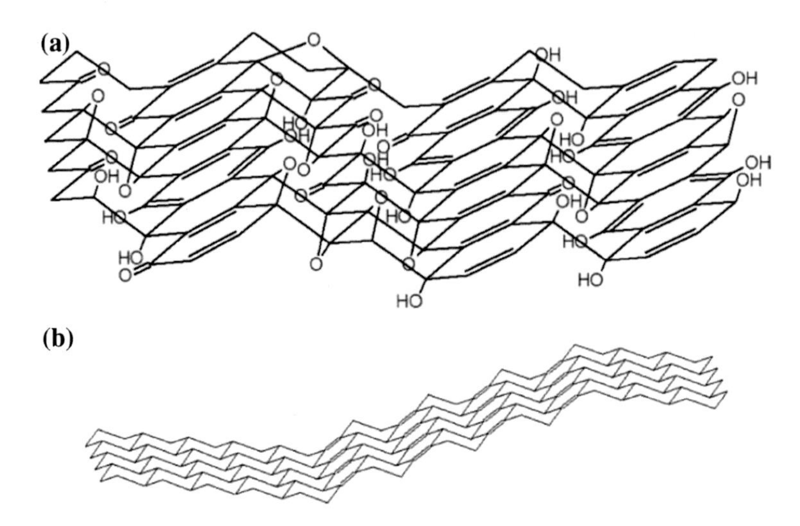

Fig. 1.7 Structure of GO proposed by Dekany and coworkers. Reprinted with the permission from Ref. [16]. Copyright 2006 American Chemical Society

The experimental observation is employed as the reference of DFT calculations, which shows that partial oxidation is more reasonable than complete oxidation in thermodynamics [34]. However, thc accurate state and distribution of oxygen-contained functional groups still strongly depend on the coverage degree on the basal plane. Theoretical prediction shows that it behaves as the ratio of epoxide to alcohol increases with the oxidation degree [34].

1.4 The Formation of Reduced Graphene Oxide

Graphite oxide and graphene oxide are insulating, as the sp^2 structure network is broken. However, the electrical conductivity can be recovered by repairing the π-network, so the reduction of graphene oxide becomes one of the most important reactions [35]. The reduction products have a series of names, such as reduced graphene oxide (rGO), chemically reduced graphene oxide (CReGO), and graphene. We name the product as 'reduced graphene oxide', because it is different from the reduced graphene in structure. These two materials are easy to be confused, even though the structures are obviously different. So these two terms should be used separately.

We call oxidized graphite as 'graphite oxide' (GO) so far. As mentioned earlier, this material contains abundant oxygen-contained functional groups (mainly hydroxyl and epoxy). Although the interlayer spacing is expanded by inserting water [23], it still maintains the stacking structure similar to graphite. We should distinguish the graphite oxide and graphene oxide when the reduction of this material is discussed. In the aspect of chemistry, graphene oxide is very similar to graphite oxide, but different. In the aspect of structure, these two materials are different. Different from the stacking structure of graphite oxide, graphene oxide can be exfoliated into single layer or few layers. Due to the hydrophilicity of graphene oxide, its surface functionalities weaken the interaction between layers, especially in alkaline medium. Many mechanical and thermal methods are used to exfoliate graphite oxide into graphene oxide, and the commonly used methods are ultrasonic dispersion or stirring of graphite oxide in water. Ultrasonic is faster than mechanical stirring in water or polar organic solvent, but the major shortcoming is the destruction of graphene oxide sheets [36]. The finally obtained nominal size of graphene oxide is reduced to hundreds of nanometers, not a few microns, and the size distribution of product is wide [37–39]. Besides, the oxidation process also breaks the structure into smaller pieces [40, 41].

The maximum dispersion of graphene oxide in solution is of great significance for further processing and application, and it mainly depends on the characteristic of solvents and the functionalization degree after oxidation (Fig. 1.8). So far, it is found that the higher the polarity, the higher the dispersion. The dispersion degree in water is generally in the magnitude of 1–4 mg mL^{-1} [42]. Ultrasound can totally exfoliate the graphite oxide, confirmed by AFM characterization of graphene oxide sheet collected from the suspension solution [37].

Reduced graphene oxide is similar to the original graphene, and the reduction is the most important reaction of graphene oxide. To the scientists and engineers who try to use graphene in large scale (such as in energy storage), the chemical conversion of graphene oxide is a realistic route. The reduction can be realized by thermal (annealing, microwave, light), chemical (reducing agent, photocatalysis,

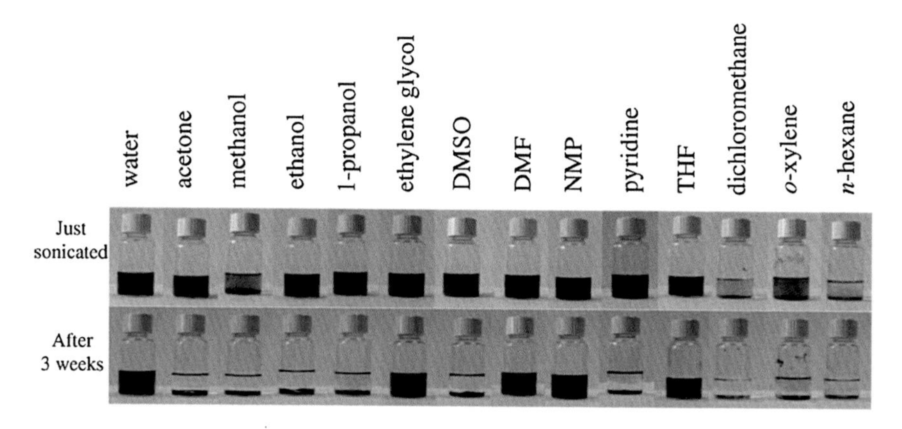

Fig. 1.8 Digital pictures of as-prepared GO dispersed in water and 13 organic solvents through ultrasonication (1 h). *Top* dispersions immediately after sonication. *Bottom* dispersions 3 weeks after sonication. The *yellow* color of the o-xylene sample is due to the solvent itself. Reprinted with the permission from Ref. [36]. Copyright 2008 American Chemical Society. Reproduced from Ref. [27] by permission of The Royal Society of Chemistry

electrochemistry, hydrothermal reaction), or combined method [35]. This dissertation will discuss two commonly used methods in detail, i.e., thermal annealing and chemical reduction.

1.4.1 Thermal Annealing

The reduction of graphene oxide can be realized by merely annealing. In the initial research of graphene, reduced graphene oxide was obtained by quick heating (>2000 °C min^{-1}) of graphite oxide [43–46]. The mechanism is that the oxygen-containing functionalities connected on the graphene surface decompose intensely during quick heating of GO, releasing CO and CO_2 between the layers. The expansion effect by gases generates huge pressure. According to the preliminary estimate, the interlayer pressure can reach 40 MPa if the temperature increases to 300 °C instantaneously, and 130 MPa to 1000 °C [45]. Calculation from Hamaker constant shows that pressure of 2.5 MPa is large enough to separate the graphene oxide sheets [45].

The exfoliated layer can be directly named as thermally reduced graphene, not graphene oxide. It demonstrates that the quick heating process not only exfoliates the graphene oxide, but also reduces the functional graphene through decomposition of oxygen-containing functional groups. This double effect renders thermally expanded graphene oxide of great potential in mass production of graphene. However, research shows that only small and wrinkled graphene sheets can be produced by this method [44]. Because the decomposition of functional groups takes away the carbon atoms on the basal plane, breaking the graphene sheet into small pieces and warping the carbon lattice (Fig. 1.9). The release of CO_2 during thermal exfoliation can severely destroy the structure of graphene [47]. The mass loss of graphite oxide is about 30 % during expansion, and amount of crystal defects are left on the whole plane [44]. The defect suppresses flexible electron transmission, and introduces amount of scattering center, which affect the electronic performance of the products inevitably. As a result, the conductivity of the obtained graphene by this method usually ranges from 10 to 23 S cm^{-1}, which is much lower than that of perfect graphene. So this method is unable to restore the electronic structure of the basal plane.

Another optional way to acquire large-sized graphene is liquid exfoliation of graphene oxide [49]. After formation of macrostructure (such as film and powder), reduction can be realized by annealing in the inert or reduction atmosphere. In this method, annealing temperature decides the reduction effect of GO [37–52]. Schniepp et al. found that the atom ratio of C/O is less than 7, if the annealing temperature is lower than 500 °C, while the ratio increased to 13 as the temperature reached 750 °C. Li et al. researched the chemical construction evolution with the changed temperature. The evolution of XPS spectrum in Fig. 1.11 demonstrated that good reduction result can be obtained at higher annealing temperature. Wang et al. [48] annealed the GO film at different temperatures, and found that the volume conductivity of

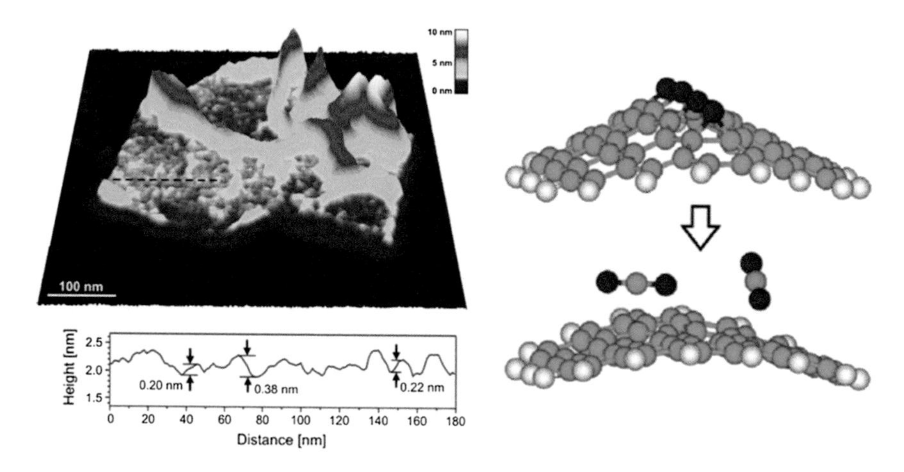

Fig. 1.9 Pseudo-3D representation of a 600 nm × 600 nm AFM scan of an individual graphene sheet showing the *wrinkled* and *rough* structure of the surface, and an atomistic model of the graphite oxide to graphene transition. Reprinted with the permission from Ref. [44]. Copyright 2006 American Chemical Society

reduced film is 50 S cm^{-1} at 500 °C, but reached 100 and 550 S cm^{-1} (Fig. 1.10) at 700 and 1100 °C, respectively. Wu et al. [52] prepared graphene by treating graphite oxide with arc discharge method. Arc discharge can generate high temperature of 2000 °C in a short time. As a result, the characteristic surface conductivity of graphene can be reached 2000 S cm^{-1}, and the atom ratio of C/O was 15–18.

Besides annealing temperature, experimental atmosphere is also important for thermal reduction of GO. The corrosion of oxygen is enhanced with the increase of temperature, so oxygen should be avoided during annealing. So the thermal

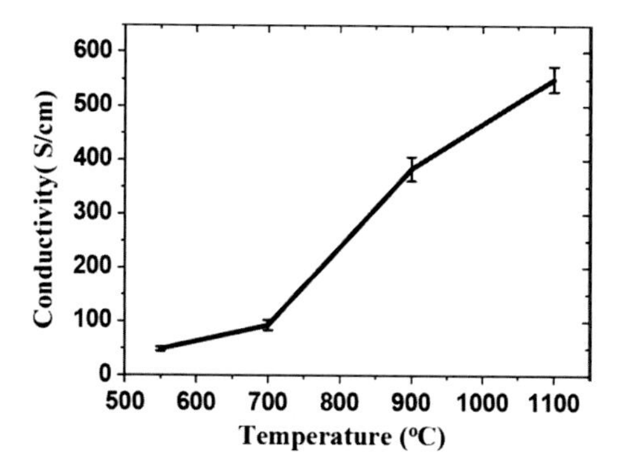

Fig. 1.10 Increase of the average conductivity of graphene films from 49, 93, 383 to 550 S cm^{-1}, along with the temperature increasing from 550, 700, 900 to 1100 °C, respectively. Reprinted with the permission from Ref. [48]. Copyright 2008 American Chemical Society

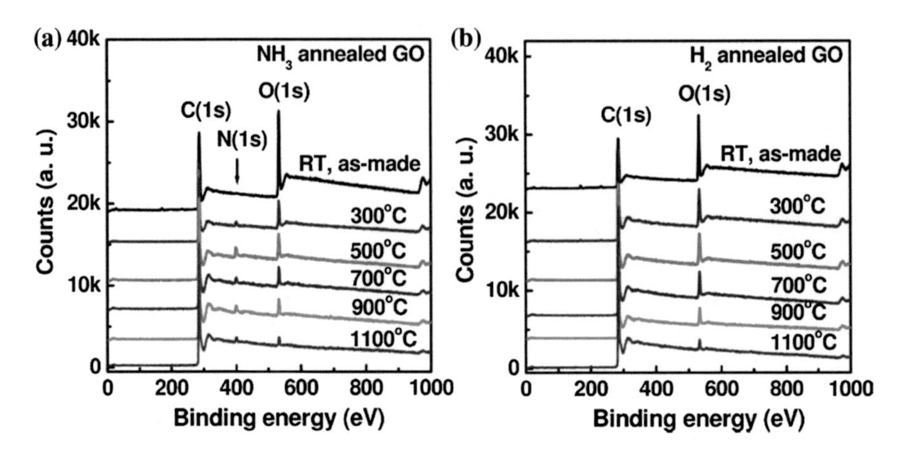

Fig. 1.11 XPS spectra of GO sheets annealed in 2 T of **a** NH_3/Ar (10 % NH_3) and **b** H_2 at various temperatures. Reprinted with the permission from Ref. [53]. Copyright 2009 American Chemical Society

reduction is carried out in vacuum [37], inert atmosphere [48], or reducion atmosphere [43, 46, 48, 53]. Becerril et al. [37] annealed the GO film at 1000 °C, and showed that high vacuum atmosphere ($<10^{-5}$ T) is essential to the restoration of GO, otherwise the film would be broken by reacting with the residual oxygen. The same situation should be considered in inert atmosphere. For example, we can consume the residual oxygen by importing hydrogen gas. More importantly, the annealing temperature in H_2 atmosphere can be properly decreased as the reduction effect is strong at annealing. Wu et al. showed that the atom ratio of C/O is 14.9, and the electrical conductivity is ~1000 S cm^{-1} after annealing GO at 450 °C for 2 h in the compound atmosphere of Ar/H_2 (1:1). Li et al. [53] report that both the reduction and *N*-doping of GO can be realized at the same time by annealing GO in NH_3 atmosphere at lower pressure (2 T NH_3/Ar (10 % NH_3)). As shown in Fig. 1.11, the N-doping content was as high as 5 % at 500 °C. The electrochemical measurement indicated that the electricity conductivity of GO reduced in NH_3 atmosphere is higher than that in H_2, and the n-type electronic doping effect was displayed clearly. This is beneficial to the further assembly of electronic devices. Recently, report showed that the surface hole could be partially repaired when rGO was exposed in the carbon source gas (like ethylene) at high temperature (800 °C). The experimental condition was similar to CVD growth of SWCNTs in fact. The surface electrical resistivity of single-layer rGO reduced to 28.6 kΩ sq^{-1} (or 350 S cm^{-1}) after the above carbon deposition [54]. Su et al. showed the similar repairing method by modifying rGO with the pyrolytic aromatic molecules. They finally obtained highly graphited material with the electricity conductivity that reached 1314 S cm^{-1} [55].

In conclusion, thermal reduction of GO is effective, but the shortcoming is obvious. First, the energy consumption of high temperature is large, and the treatment condition is rigorous. Second, the heating speed must be slow enough to prevent expansion during the reduction of assembled GO macrostructure (such as GO film), otherwise the structure would be destroyed by quickly increased temperature, which is similar to the thermal expansion of GO. But it is time consuming to anneal GO at slowly increased temperature. Last but most important, some applications need to assembly GO on the substrate (such as super thin carbon film). High temperature means that this reduction method is inappropriate for reducing GO film deposited on substrate with low melting point (such as glass and polymer).

1.4.2 Chemical Reduction

Reduction by chemical reagent is based on the chemical reaction with GO. Generally, the reduction can be realized at room temperature or mild heating condition. Therefore, the equipment and environment requirements of chemical reduction are not as rigorous as thermal annealing. So compared with latter, the former is cost effective and operable to prepare graphene in large scale.

Many methods can be used to characterize the reduced products and reactants. First, the atom ratio of C/O is an important parameter to evaluate the reduction degree. Second, BET specific surface area can be used. Simply speaking, quantitative analysis of the specific surface area is carried out by calculating the physical adsorption amount of gas (normally nitrogen) on the material surface [56]. Raman spectroscopy is also important in evaluating the reduction effect, and the major vibration peaks of graphite materials are D peak (relate with ordered/disordered states of the system) and G peak (relate with stacking structure). The peak intensity ratio of D/G is used to measure the layers and stacking behavior of graphene. Higher D/G ratio means higher exfoliation ratio and disordered degree [57]. The electrical property is certainly meaningful, generally expressed by volume electrical conductivity (σ, S m^{-1}), as well as other reported parameters like surface resistivity (R_{sh}, Ω sq^{-1}) or surface electrical conductivity (G_{sh}, S sq^{-1}). Surface resistivity is the resistive performance of surface, irrelevant with the material thickness. Its relationship with the volume electrical conductivity is: $R_{sh} = 1/\sigma t$, among which t is the thickness of material. Other methods are also used, including atomic force microscopy (AFM), X-ray photoelectron spectroscopy (XPS), scanning electron microscopy (SEM), and transmission electron microscopy (TEM) [58, 59].

Hydrazine was used to reduce graphite oxide [61] before the discovery of graphene, but Stankovich et al. first reported that it can also be used to reduce exfoliated graphene oxide [60, 62] (Fig. 1.12). These reports opened the door to large-scale production of graphene. After that, hydrazine was considered to be an effective chemical agent to reduced GO [38, 50, 62–74]. This process only needs to add hydrazine or its derivatives (such as hydrazine hydrate, dimethylhydrazine) into GO aqueous solution, and then the aggregated graphene nanosheets are obtained. The aggregation is ascribed to the enhanced superficial hydrophobicity of

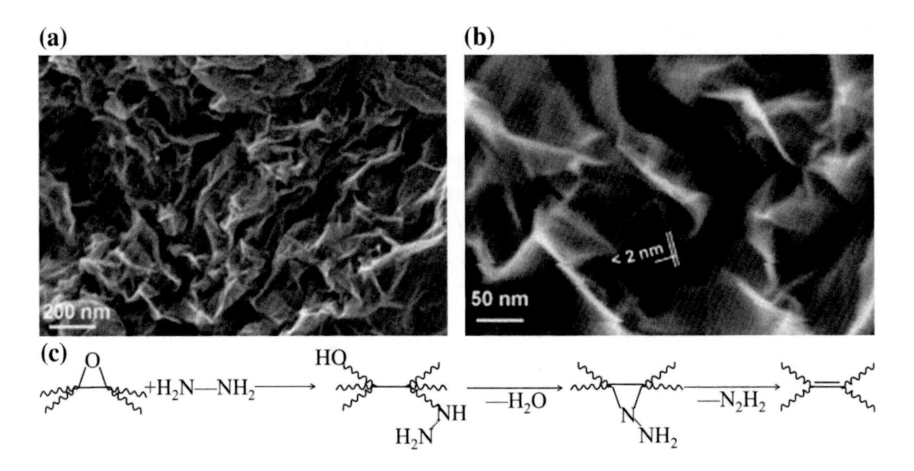

Fig. 1.12 **a** An SEM image of aggregated reduced graphene oxide platelets; **b** A platelet having an *upper bound* thickness at a fold of 2 nm; **c** A proposed reaction pathway for epoxide reduction by hydrazine. Reprinted from Ref. [60], Copyright 2007, with permission from Elsevier. Reproduced from Ref. [27] by permission of The Royal Society of Chemistry

the reduced material. The black conductive powders are obtained after drying the above products, with C/O ratio approaching to 10 [60]. The electrical conductivity was as high as 99.6 S cm^{-1} by reducing GO film with hydrazine, and the corresponded C/O ratio was about 12.5 [64]. In the application of graphene, it is necessary to maintain the good dispersion in water during the reduction of GO, and this can be achieved by adding soluble polymer as surfactant [62], or modulating surface electric potential of rGO with ammonium hydroxide [63]. Well-dispersed graphene sheets in colloidal solution can be assembled into bulk through easily solution process (such as filtration) [63].

Metal hydroxides (such as NaH, $NaBH_4$, and $LiAlH_3$) are commonly used as strong reducing agent in organic chemistry, but these agents are more or less active with water, which is the main solvent in dispersion and exfoliation of GO. $NaBH_4$ is recently proved to be more effective than hydrazine in reducing GO [75], even though it is hydrolyzed slowly. Results showed that the dynamic speed of hydrolyze was so low that the newly prepared solution could be used to reduce GO. $NaBH_4$ is effective in reducing C=O group [76], but poor in epoxy and carboxyl groups. So the hydroxyl group still existed after reducing. Gao et al. [31] made some improvements to enhance the reducing effect, proposed to add dehydration process by concentrated sulfuric acid at 180 °C after reducing by $NaBH_4$. The obtained C/O ratio of rGO was about 8.6, and the electrical conductivity was about 16.6 S cm^{-1} by this two-step method (Table 1.2).

Table 1.2 Comparison of the reducing effect of GO by different methods

Reduction method	Form	C/O ratio	σ (S cm^{-1})
Hydrazine hydrate	Powder	10.3	2
Hydrazine reduction in colloid state	Film	NA[b]	72
150 mM $NaBH_4$ solution, 2 h	TCF	8.6	0.045
Hydrazine vapor	Film	~8.8	NG
Thermal annealing at 900 °C, UHV[a]		~14.1	NG
Thermal annealing at 1100 °C, UHV[a]	TCF	NA	~10^3
Thermal annealing at 1100 °C in Ar/H_2	TCF	NA	727
Multistep treatment			
(I) $NaBH_4$ solution	Powder	(I) 4.78	(I) 0.823
(II) Concentrated H_2SO_4 180 °C, 12 h		(II) 8.57	(I) 16.6
(III) Thermal annealing at 1100 °C in Ar/H_2		(III) > 246	(III) 202
Vitamin C	Film	12.5	77
Hydrazine monohydrate		12.5	99.6
Pyrogallol		NA	4.8
KOH		NA	1.9×10^{-3}
55 % HI reduction	Film	>14.9	298

[a]*UHV* ultra high vacuum
[b]*NA* not available

Ascorbic acid (vitamin C, VC) is a recently reported reducing agent for GO, and supposed to be the ideal replacement of hydrazine [64]. Fernandez-Merino et al. reported that the C/O ratio of VC reduced GO was about 12.5, electrical conductivity was 77 S cm^{-1}, which were comparable with those by hydrazine. Moreover, the greatest advantage of VC is nontoxic compared with hydrazine, and chemically stable in water compared with $NaBH_4$. It can also avoid the aggregation in colloidal suspension appeared in hydrazine reduction, which facilitates the further applications.

Recently, Pei [77] and Moon et al. [78] reported another strong reducing agent, hydroiodic acid (HI). These two separate researches reported similar reducing effects: the C/O atom ratio of rGO was about 15, and the electrical conductivity of rGO film was about 300 S cm^{-1}, which were higher than those reduced by other chemical methods. HI can reduce GO in different states, including colloid, powder, or film, in both vapor and liquid phases, even at room temperature [78]. Figure 1.13 showed the optical photographs of GO films chemically reduced by different agents (HI, hydrazine vapor, 85 % N_2H_4/H_2O, $NaBH_4$ solution). GO film reduced by HI possesses good flexibility and increased tensile strength, but the film reduced by hydrazine vapor is hard and difficult to bend with its thickness increasing more than 10 times. Correspondingly, films reduced by hydrazine hydrate or $NaBH_4$ cannot maintain their morphology and collapse into fragments. Results showed that reducing effect of HI was better than that by hydrazine hydrate, and the former was more suitable for reducing GO film. So HI can not only be used to reduce GO film but also obtain TCFs with high performance [49, 77].

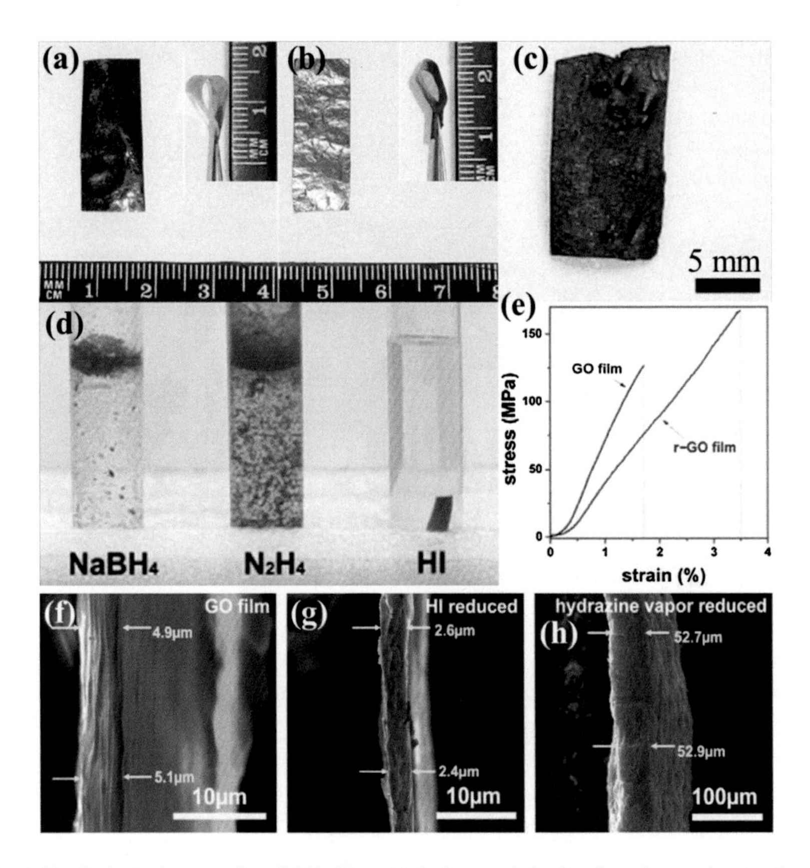

Fig. 1.13 Optical photographs of GO films (**a**) before and (**b**–**d**) after chemical reduction by different agents: **b** HI, **c** hydrazine vapor, **d** 85 % N_2H_4/H_2O (N_2H_4), 50 mM $NaBH_4$ solution ($NaBH_4$) and 55 % HI after immersion for 16 h at room temperature, **e** the stress–strain curve of the GO film and HI-reduced GO film (r = stress, e = strain), and SEM images of the cross-sectional views of GO films (**f**) before and (**g**, **h**) after reduction by (**g**) HI and (**h**) hydrazine vapor. Reprinted from Ref. [77], Copyright 2010, with permission from Elsevier

Some other reducing agents can also be used to reduce GO, such as hydroquinone [79], biphenyl three phenol [64], hot strong alkali solution (KOH, NaOH) [80], hydroxylamine [81], carbamide, and thiocarbamide. But compared with strong reducing agents (such as previously reported hydrazine, $NaBH_4$, HI, etc.), the reducing effects by these agents are somewhat poor.

1.5 Application Prospect of Reduced Graphene

As a rising star in nanocarbon material, graphene brings the research of 'charcoal' and 'carbon' into new age. Carbon atoms in graphene hybridize in sp^2 honeycomb, and the surface exposes to the external environment in the most opened manner, which is different from the traditional carbon materials (such as graphite, carbon fiber, active carbon, glassy carbon, carbon black, etc.) and nanocarbon materials (such as fullerene, carbon nanotube, nanodiamond, etc.). Both the plane and edges are decorated with abundant active sites, including oxygen-containing functional groups (such as hydroxyl, carboxyl, carbonyl, epoxy groups, etc.) and crystal defects (such as vacancy, distortion, dangling bond, etc.), especially for the chemically prepared graphene. Introducing the functional group, on the one hand, enhances the surface wettability and interfacial compatibility of graphene, which provides opportunity for bulk assembly in wet environment; on the other hand, the original active sites can hybridize with N, P, and B, which can further modulate surface alkalinity/acidity and electron donor/receptor performance. These special advantages bring graphene with abundant surface/interface performances, laying foundation for promising applications in energy storage and conversation, catalyst, composite material, etc. (Fig. 1.14). In the following, we will mainly discuss the applications of graphene in paper/membrane-like materials and electrochemical energy storage.

1.5.1 Paper/Membrane-Like Materials

1.5.1.1 Graphite Oxide Paper

Directional filtration of water-dispersed layered clay (such as vermiculite and mica) can obtain free-standing paper-like materials, and this technology has been commercialized successfully [82]. Dikin et al. used the above technology into dispersion system of graphene oxide, and prepared graphene oxide paper (Fig. 1.15b) [83]. This dark paper-like material possessed obvious layered structure with interlayer spacing of 8.3 Å, which was closed to the non-exfoliated GO (6.8 Å) [87]. The interlayer spacing was mainly affected by intercalated water [83]. As shown in SEM images (Fig. 1.15c), graphene oxide paper was constructed by stacked layers closely. These layers were interlocked with each other, showing wavy character on the surface. The thickness of the graphene oxide paper (1–30 μm) can be controlled

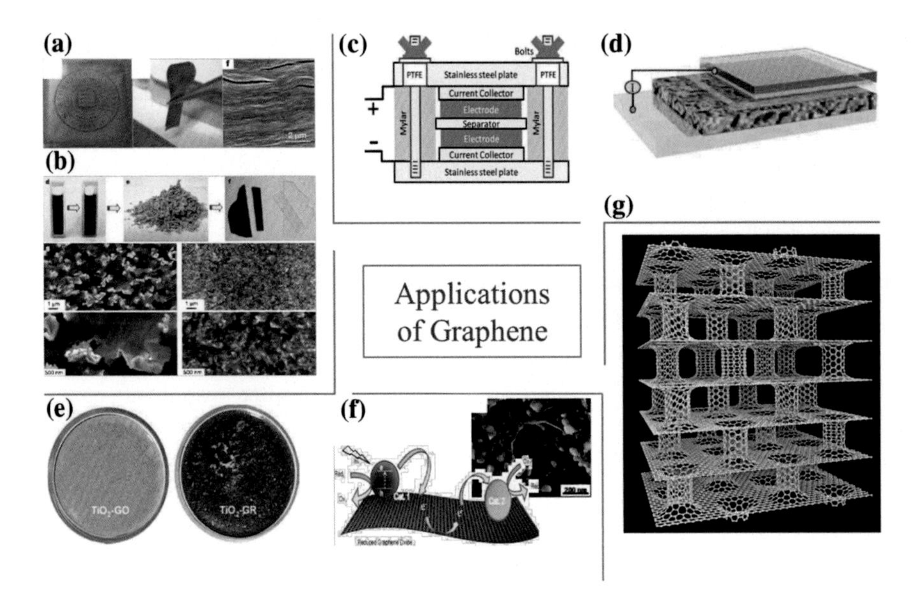

Fig. 1.14 Promising applications of chemically derived graphene. **a** Reprinted by permission from Macmillan Publishers Ltd.: Ref. [83], copyright 2007. **b** Reprinted by permission from Macmillan Publishers Ltd.: Ref. [39], copyright 2006. **c** Reprinted with the permission from Ref. [120]. Copyright 2008 American Chemical Society. **d** Reprinted with the permission from Ref. [48]. Copyright 2008 American Chemical Society. **e** Reprinted with the permission from Ref. [84]. Copyright 2008 American Chemical Society. **f** Reprinted with the permission from Ref. [85]. Copyright 2010 American Chemical Society. **g** Reprinted with the permission from Ref. [86]. Copyright 2008 American Chemical Society

by changing graphene oxide content in the filtration system. The paper was semitransparent in lower thickness (<5 μm), but nontransparent in larger thickness [83]. Almost completely transparent paper (transmittance >90 %) can also be prepared through filtration method with thickness of about 2 nm [88].

The graphene oxide paper is nonconducting [68], but exhibits good mechanical strength (Table 1.3) with Young's modulus as high as 23–42 GPa [82, 92, 93], which is comparable to concrete [91], but much higher than that for inorganic vermiculite paper (14.1 GPa) [82] or CNT self-assembled Bucky paper (2.7 GPa) [94]. The ultimate tensile strength of graphene oxide paper is 15–193 MPa [82, 92, 93], comparable to cast iron [95] and vermiculite paper (160 MPa), but much higher than that of Bucky paper (33.2 MPa). These mechanical performances can be improved by decorating functional groups on graphene oxide sheets. For example, in the in situ preparation of graphene oxide paper, Young's Modulus increases 10–40 %, and tensile strength 10–80 % by cross-linking the neighboring carboxyl groups on the edges with 1 wt% bivalent cation (Mg^{2+} or Ca^{2+}) [93]. Amine-functionalized graphene oxide paper was obtained by treating the graphene oxide paper with ethylamine, which opened the epoxy groups on the basic plane [92]. Ethyl group filled

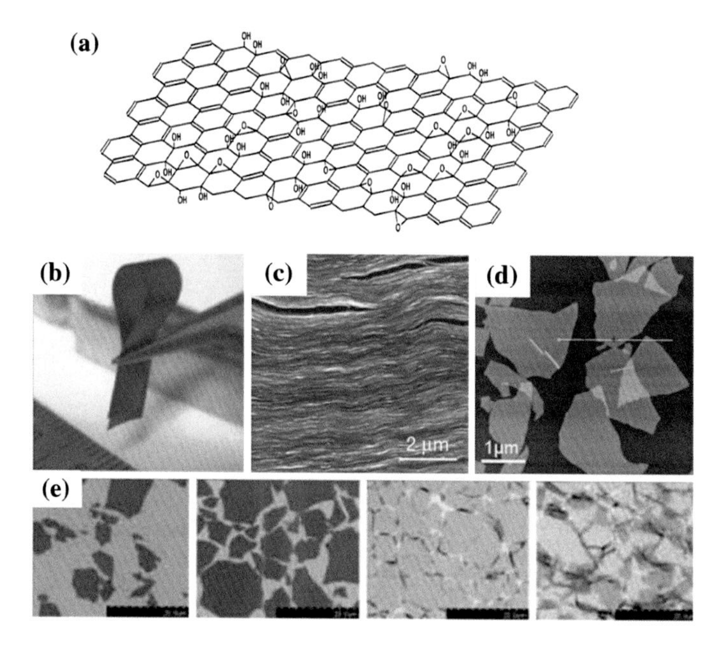

Fig. 1.15 **a** Schematic model of a graphene oxide sheet. Reprinted from Ref. [28], Copyright 1998, with permission from Elsevier. **b** Photograph of a graphene oxide paper ribbon, and **c** SEM image of the edge of graphene oxide paper showing tightly packed, undulating structure. Reprinted by permission from Macmillan Publishers Ltd.: Ref. [83], copyright 2007. **d** A tapping mode AFM image of graphene oxide sheets on mica surface. Reproduced from Ref. [89] by permission of The Royal Society of Chemistry. **e** SEM images of a Langmuir–Blodgett-assembled graphene oxide thin film collected on a silica substrate with surface pressure increasing from left to right. Reprinted with the permission from Ref. [90]. Copyright 2009 American Chemical Society

into the layer gap, and stabilized the structure after annealing at 300 °C. 68 % tensile strength and 35 % Young's modulus were retained after annealing. It formed striking contrast with unmodified graphene oxide paper, which totally lost mechanical strength after annealing.

Graphene oxide paper can also be prepared by evaporating water in the dispersion system, and the self-assembly process takes place during the evaporation [96]. The interlayer spacing of this paper is 7.5 Å, which is between the value of directional filtrated graphene oxide paper (8.3 Å) [82] and graphene oxide powder (6.8 Å) [87]. The thickness of self-assembled paper by evaporation (0.5–20 μm) can be controlled by changing the concentration of dispersion system. The layers stack compactly in the self-assembled paper, but the corresponding mechanical performances (Young's modulus ≈ 12.7 MPa, tensile strength ≈ 70 MPa) are worse than that of directional filtrated paper (Table 1.3).

Table 1.3 Mechanical performances of self-assembled graphene and graphene oxide paper

Type of paper	Average Young's modulus [GPa]	Average ultimate tensile strength [MPa]
Graphene oxide paper	32	130
Graphene oxide paper, annealed (300 °C)	Not determinable	Not determinable
Graphene oxide paper, hexylamine-modified	17.2	63.4
Graphene oxide paper, hexylamine-modified, annealed (300 °C)	5.9	42.8
Graphene paper, flow-directed assembled	20.1	152
Graphene paper, reduced in situ	3.0	12.7
Graphene paper, flow-directed assembled, annealed (220 °C)	41.8	293.3
Graphene paper, reduced in situ, annealed (300 °C)	3.4	25.0
Graphene oxide paper	25.6	81.9
Graphene oxide paper, Mg^{2+}-modified	27.9	80.6
Graphene oxide paper, Ca^{2+}-modified	21.8	125.8
Graphene oxide membrane, self-assembled	12.7	70.0
Bucky paper	2.7	33.2

Reproduced from Ref. [13] by permission of John Wiley & Sons Ltd.

1.5.1.2 Electric-Conductive Reduced Graphene Paper

Similar to directional-filtrated graphene oxide paper, filtration of reduced graphene dispersion system can also obtain self-supporting paper-like material (Fig. 1.16c and d) [63, 68, 97]. The degree of order for this kind of paper is poor, having a weak and broad diffraction peak at 3.9 Å. Correspondingly, the mechanical strength is poor (Young's modulus = 20.5 GPa, tensile strength = 150 MPa), and the in-plane electricity is low (≈7200 S cm^{-1}, measured by four probe method). The reduced graphene paper tends to be ordered after annealing at 500 °C, and a sharp diffraction peak appears at 3.4 Å (3.35 Å for original graphite). The electricity of reduced graphene paper after annealing is 35,000 S m^{-1}, the largest value ever since reported for self-supported reduced graphene-based materials. Moreover, the mechanical strength is improved after annealing at 220 °C, with Young's modulus of 41.8 GPa and tensile strength of 293.3 GPa. But these values decrease after annealing at higher temperatures. The mechanical strengths of graphene oxide paper

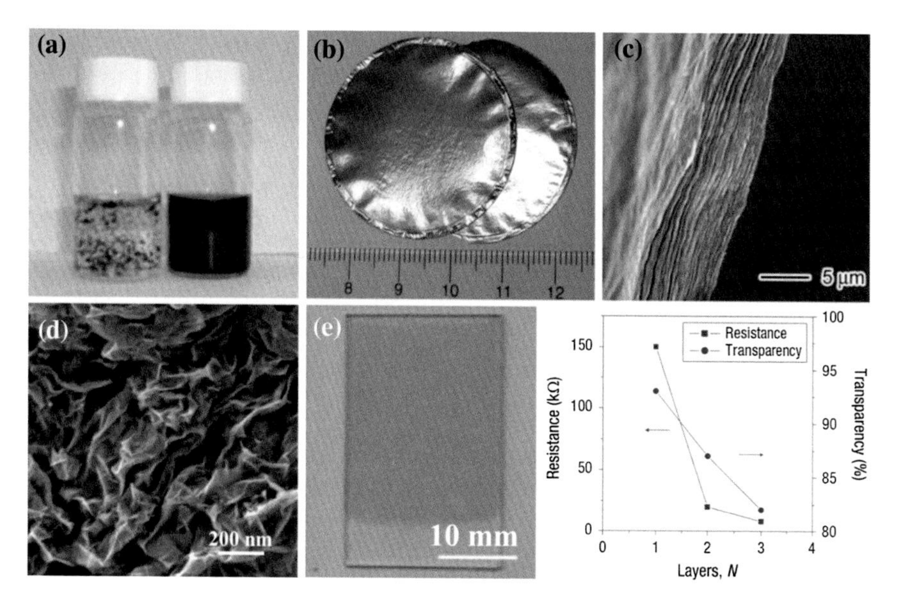

Fig. 1.16 **a** Photograph of aqueous solutions of graphene sheets reduced in the absence (*left*) and presence (*right*) of poly (sodium4–styrenesulfonate) (PSS), demonstrating the critical role of the amphiphilic PSS surfactant in maintaining a dispersion after graphene oxide reduction. Reproduced from Ref. [62] by permission of The Royal Society of Chemistry. **b** Photograph of graphene paper produced by filtration of aqueous graphene dispersion. Reproduced from Ref. [68] by permission of John Wiley & Sons Ltd. **c** SEM image of the layered structure of a graphene paper, viewed from a fracture edge. Reproduced from Ref. [68] by permission of John Wiley & Sons Ltd. **d** SEM image of aggregated graphene sheets in the form of graphene powder. Reprinted from Ref. [60], Copyright 2007, with permission from Elsevier. **e** Photograph of a transparent, conductive graphene thin film fabricated via Langmuir–Blodgett assembly with plot correlating decreased resistance and transparency with increased film thickness. Reprinted by permission from Macmillan Publishers Ltd.: Ref. [99], copyright 2008

decrease after annealing, and become corrugated after annealing at 300 °C, which form striking contrast with reduced graphene paper. It is ascribed to the fact that oxygen-contain functional groups decompose and form plenty of gas when the annealing temperature is higher than 150 °C [98]. The structure of graphene oxide paper is destroyed after annealing, resulting in decreased mechanical performance.

In addition, reduced graphene paper can be obtained by reducing graphene oxide paper, and the present technology includes in situ hydrazine [92] and flash [90] reduction. In situ hydrazine reducing is similar to the process of reducing graphene oxide dispersion system by hydrazine. In this process, graphene oxide paper supported by filter membrane is employed to filtrate the hydrazine aqueous solution [92]. But the obtained paper is not uniform partially, resulting in decreased in-plane electric conductivity (≈200 S m^{-1}) and mechanical strength (Young's modulus = 3.0 GPa, tensile strength = 13.2 MPa). As for directional-filtrated reduced graphene paper, these performances (electric conductivity = 1700 S m^{-1}, Young's modulus = 3.4 GPa, tensile strength = 24.7 MPa) are improved after annealing at 300 °C. Considering the simplicity, in situ reducing method is used to prepare chemically modified reduced graphene paper from modified graphene oxide paper [92].

Flash technology needs to expose graphene oxide paper under xenon lamp equipped in digital camera (energy region is normally between 0.1–0.2 J cm^{-2}). This optothermal reducing method increases the thickness of paper (1 μm) with two orders of magnitude, resulting from decomposition and escape of gas at high temperature. The obtained material is fluffy with, respectively, low electric conductivity (1000 S m^{-1}). Considering the intrinsic quality of this flash-motivate reducing method, staggered electrode can be prepared, with strip-like electric pattern on the surface, by covering shelter on the original surface before reduction [90].

The existence of oxygen-containing functional groups on the surface has negative effects on the conductivity of bulk materials fabricated from reduced graphene. The electric conductivities of reduced graphene paper are 7200 and 1690 S cm^{-1} for those fabricated from aqueous [68] and organic [100] solutions, respectively, but only 687 and 200 S m^{-1} for those using K^+ [101] and 1-pyrenebutyrate [102] as stabilizers. Difference is obvious between the above two methods. It should be noted that electric conductivity can be increased by decreasing the interlayer spacing of reduced graphene paper, and the simplest method is annealing. Electron transfer between layers is enhanced by decreasing interlayer spacing. For example, the electric conductivity increases by one order of magnitude when the interlayer spacing decreases from 3.86 to 3.7 Å [100]. Interestingly, the electric conductivity of reduced graphene paper is increased after annealing at higher temperature, but there is no direct relationship between interlayer spacing and electric conductivity (Fig. 1.17).

El-Kady et al. [103] published a paper in *Science* recently, pointing out that in the standard LightScribe DVD drive, graphene oxide film deposited on the media disk is reduced into porous graphene film by laser irradiation. The obtained film has relatively high mechanical strength, with electric conductivity of 1738 S cm^{-1}, specific surface area of 1520 m^2 g^{-1}. This film can be directly used as flexible electrode in electrochemical capacitor without binder or current collector. The fabricated flexible energy storage device has high power density and energy density, showing wide application prospect in flexible electronic devices with high powder density.

1.5.1.3 Transparent Conducting Film

Graphene has excellent electron transfer character and high carrier concentration, and its light transmittance of single layer is higher than 97 %. As an ideal two-dimensional material, graphene can be assembled into film electrode with high transmittance, high electric conductivity, low roughness, and good flexibility, which is supposed to replace expensive fragile ITO (nano-Indium Tin Oxides) and FTO (Fluorine-doped Tin oxide) glass (Fig. 1.18). One of the most promising applications for graphene-based materials is transparent conductive film (Fig. 1.16e). This film

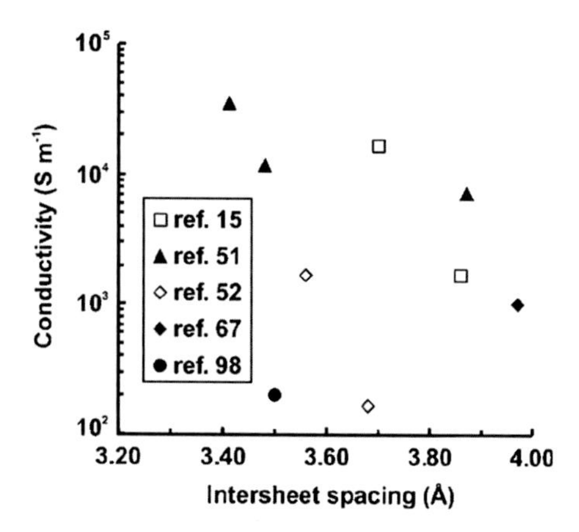

Fig. 1.17 The relationship between in-plane conductivity and interlayer space of free-standing graphene paper. Reproduced from Ref. [13] by permission of John Wiley & Sons Ltd.

can be used as the electrode of solar cell, transistor of next generation of displayer, high sensitive detector for explosives. As discussed in the first part, mechanical exfoliation [4], vapor deposition [104, 105], and epitaxial growth [106] can be used to prepare highly oriented single-layer graphene. But only large area film has broad applications. The fabrication method is still immature. It is easy to fabricate film of large area from graphene/reduced graphene dispersion systems or reducing graphene oxide film, but some inherent defects, suppressing electric conductivity, are generated in these films during oxidation–reduction process (Fig. 1.15a). So this approach needs to be further studied to prepare large area film with high conductivity comparable with graphene [44, 60].

At present, the most convenient and efficient way to prepare large area graphene film is to reduce the previously assembled graphene oxide film. Reduction includes heat annealing [48], exposing in hydrazine vapor [90], or combining the above two methods (Table 1.3) [37, 108]. Becerril et al. [37] obtained highly reduced graphene film by annealing graphene oxide film at 1100 °C, and found that the layered resistivity decreased by four orders of magnitude compared with that exposed in hydrazine vapor. Therefore, high temperature is important for preparing film with low-layered resistivity. For example, the resistivity of film (10 nm in thickness) is 2 kΩ after annealing at 900–1000 °C, but 15 kΩ at 550–700 °C. The former is one magnitude less than the latter [48]. In addition, the transmittance of light with wavelength larger than 300 nm is increased after high temperature annealing, and this is because the degree of graphitization is increased. The first graphene paper in the world is prepared by this method, with thickness of 10 nm, transmittance higher than 70 %, and electric conductivity of 550 S cm^{-1}, and it has been first used in dye sensitization solar cell. Its photoelectric conversion efficiency (η) is only 0.26 %, and short circuit current is 1.01 mA cm^{-1}. This is because plenty of grain boundaries and defects are formed in the film during wet assembly of functionalized graphene, resulting in enhanced internal resistance. Mild intercalation of graphite can be considered to improve the conductivity of film, and maintain high transmittance at the same time.

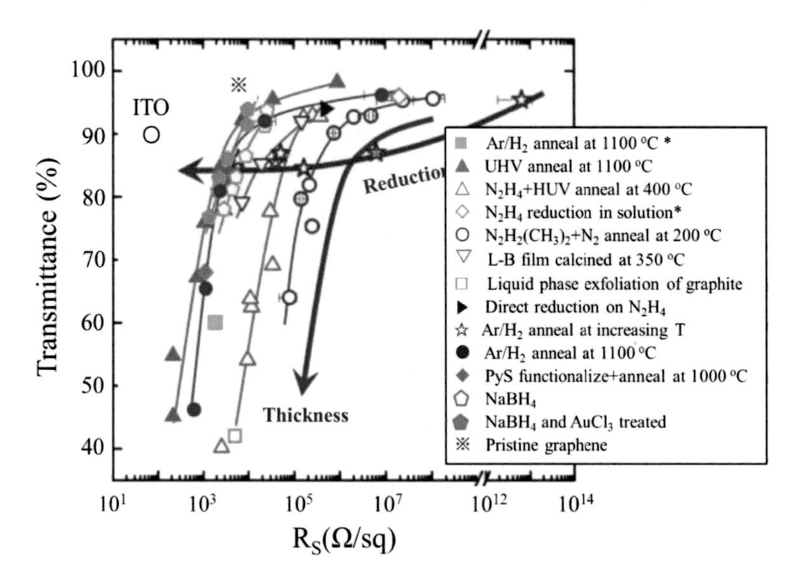

Fig. 1.18 Transmittance at 50 nm as a function of sheet resistance of rGO films reported by various groups in the literature. Reproduced from Ref. [2] by permission of John Wiley & Sons Ltd. In some cases, indicated by *asterisk*, the properties were estimated from the information provided. The *stars* represent the values for the same film thickness annealed at different temperatures. The *asterisk* in the plot corresponds to the values for ideal graphene, whose transmittance and sheet resistance are 98 % [107] and 6.45 kV [4], respectively

The finally obtained graphene sheet is in large area with low defects, which can be assembled in nanoscale. This method is likely to be effective in fabricating large-area graphene. At the same time, graphene can be grown on the substrate by chemical vapor deposition, and this has acquired good effect in transferring to solar cell. Figure 1.19 is the schematic of graphene electrode in solar cell.

Dai et al. compared the graphene film obtained by annealing LB self-assembled graphene oxide film and the reduced graphene film obtained by LB self-assembly from reduced graphene dispersion system [99]. The layered resistivity of the former is about three magnitudes higher than that of the latter. This may be ascribed to the residual oxygen-containing functional groups on the surface of reduced graphene oxide film. More electron transfer path is produced when the film thickness increases from 1 to 3 nm, resulting in obviously decreased layered resistivity from 150 to 8 kΩ, but the transmittance is decreased (from 93 to 83 % at thickness of 1000 nm).

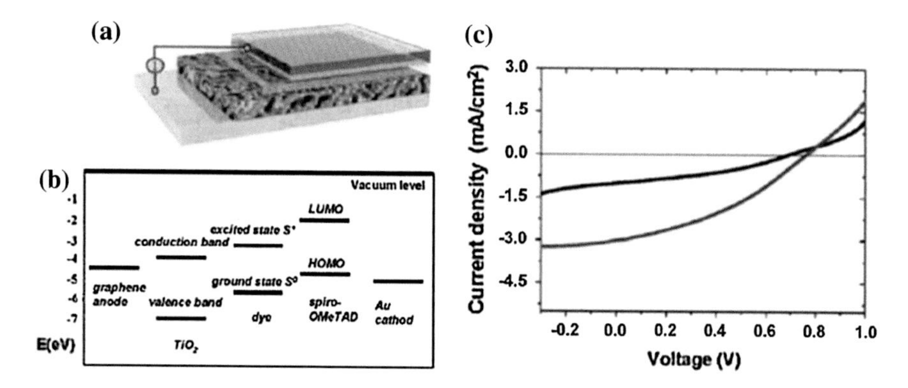

Fig. 1.19 **a** Schematic illustration of DSSC using rGO as the window electrode. **b** The energy-level diagram of DSSC consisting of GO/TiO_2/dye/spiro-OMeTAD/Au layers. **c** I/V characteristics of rGO-based (*black*) and FTO-based (*red*) DSSC under simulated solar illumination. Reprinted with the permission from Ref. [48]. Copyright 2008 American Chemical Society

1.5.2 Energy Storage Materials

1.5.2.1 Supercapacitor

Supercapacitor employs surface energy storage technology and stores electric energy directly by forming electric double layer (EDL) between material's surface and electrolyte. It can provide high power and current density in a short charge and discharge cycle. The energy storage mechanism is based on EDL coming from accumulated static charges on the interface between electrode and electrolyte and pseudocapacitance coming from quick and reversible surface oxidation–reduction reaction in the specific electric potential. The structure of supercapacitor is similar to battery, composed of two pieces of electrodes which is contacted with electrolyte, but separated by diaphragm (Fig. 1.20a). The overall performance of the capacitor is affected by every component, including electrode, separator, current collector, and electrolyte. All these factors should be considered in designing capacitor. Porous carbon is commonly used as electrode material due to its good conductivity and high specific surface area. Figure 1.20b showed the structure of EDL which is filled with positive charge on the surface of porous electrode (Stern model, ignoring the curvature of hole). The negative charges on both Stern layer and diffusion layer contribute to EDL.

As for EDL supercapacitor, every specific capacitance (F g^{-1}) on the electrode surface is supposed to follow the rule of paralleled double plate capacitor [109]:

$$C = \frac{\varepsilon_r \varepsilon_0}{d} A \tag{1.3}$$

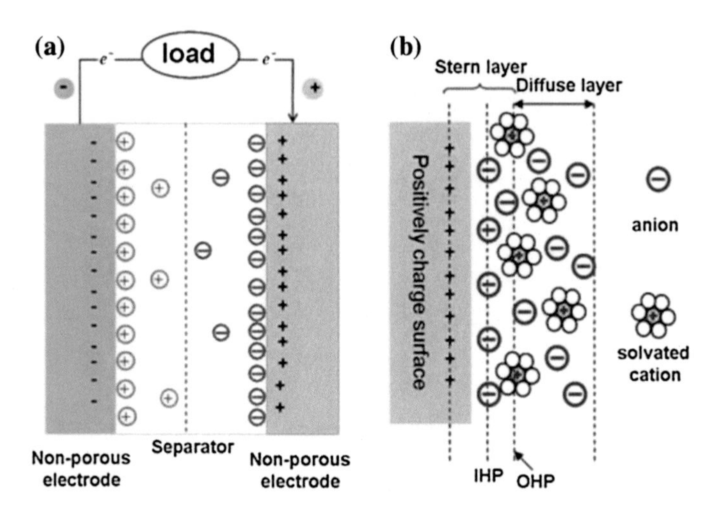

Fig. 1.20 Schematic diagrams of **a** a two-cell supercapacitor device made of nonporous electrode and **b** the EDL structure based at a positively charged electrode surface. Reproduced from Ref. [110] by permission of The Royal Society of Chemistry

where ε_r (dimensionless constant) is the relative permittivity, ε_0 (F g^{-1}) is the permittivity of free space, A (m^2 g^{-1}) is the specific surface area of electrode by the electrolyte accessibility, d (m) is the effective thickness of EDL (also known as the Debye length). Different from EDL, pseudocapacitance is decided by the thermodynamic balance between the amount of the charge acceptance (Δq) and the change of voltage (ΔV) [111]. So the pseudocapacitance is deduced by $C = \mathrm{d}(\Delta q)/(\mathrm{d}\Delta V)$. The nature of pseudocapacitance is the Faradaic process depending on quick and reversible redox reactions between electrolyte and electrochemical active sites on the electrode surface, which is completely different EDL capacitance. The most commonly used active materials include RuO_2 [112], MnO_x [113, 114], VN [115], conducting polymer [116, 117], and the surface functional groups of carbon materials [118, 119]. Pseudocapacitance can be much higher than EDL capacitance, but the shortcomings of the former are low power density and poor cycling stability.

The performances of supercapacitor are evaluated mainly based on the following standards: (1) the overall power density is larger than that of battery, and the energy density is large (>10 Wh kg^{-1}); (2) excellent cycling stability (larger than 100 times of battery); (3) quick charge/discharge (within several seconds); (4) low self-discharge rate; (5) safety; and (6) low cost. It should be pointed out that the time constant (i.e., the product of electrical resistance (Ω) and capacitance (C)) is another important parameter evaluating the performance of supercapacitor. Therefore, in order to reduce electric leakage or self-discharge, large time constant is needed.

The maximum stored energy and discharge power for a supercapacitor unit can be calculated using the following equations:

$$E = \frac{1}{2} C_T V^2 \tag{1.4}$$

$$P = \frac{V^2}{4R_s} \tag{1.5}$$

where V (V) is the maximum cell voltage, C_T (F) is the total capacitance, R_s (Ω) is the equivalent series resistance (ESR). The above parameters are important for the performance of supercapacitor. The unit capacitance depends on the electrode materials, and the voltage is limited by the thermodynamic stability of electrolyte. ESR derives from various resistances, including the nature electrical performance of electrode and electrolyte, interfacial charge-transfer resistance, contact resistance between current collector and electrode, etc. Therefore, large capacitance, wide voltage, and low ESR must be satisfied at the same time for a supercapacitor with high performance. In the other word, the development of electrode materials and electrolyte is necessary way to optimize the performance of capacitor.

For the electrode of supercapacitors, large specific surface area and reasonable pore structure are required. Besides, the conductivity and stability of the electrode must be high. If necessary, the addition of pseudocapacitance active material also greatly increases the overall capacitance. Furthermore, the density of electrode must be large enough to provide higher volumetric energy density, which is helpful for design device with high energy and power density. On the other hand, the electrolyte must be carefully selected and evaluated. Nonaqueous electrolyte with low resistance is a good choice for design supercapacitors with high power and energy density, because of its higher work voltages (up to 3.5–4 V).

The 'opened' 2D structure of graphene can greatly improve the utilization efficiency of carbon atoms, which is consistent with the surface energy storage mechanism of supercapacitors. EDLC is in proportion to the specific surface area, and in aqueous electrolyte the surface specific capacitance of 'carbon' is 0.1–0.2 F m^{-2}, while the theoretical specific surface area of graphene is 2630 m^2 g^{-1}, so its capacitance can be as high as 263–526 F g^{-1}, which is much higher than that of the current commercial activated electrode (usually less than 200 F g^{-1}) [110]. The 2D surface of graphene can fully contact with electrolyte, and the macroporous structure between its layers provides sufficient buffer space and smooth tunnel for migration of electrolyte ions. In addition, the conjugated π electrons (high density carrier) in plane of graphene can provide low impedance paths for charge transport, satisfied the energy storage requirement of rapid charge/discharge with large current.

Ruoff et al. [120] first found that chemically modified graphene (CMG) could show higher capacitance, and the capacitance can be 135 and 99 F g^{-1} in water and organic systems, respectively (Fig. 1.21). After further improvement by Chen et al. the capacitance of graphene-based material can be increased to 205 F g^{-1} [121]. The capacitance varied from different researchers' report, because the chemically reduced graphene tends to aggregate again to form a graphite-like structure. So the practical specific surface area is much lower than the maximum theoretical value,

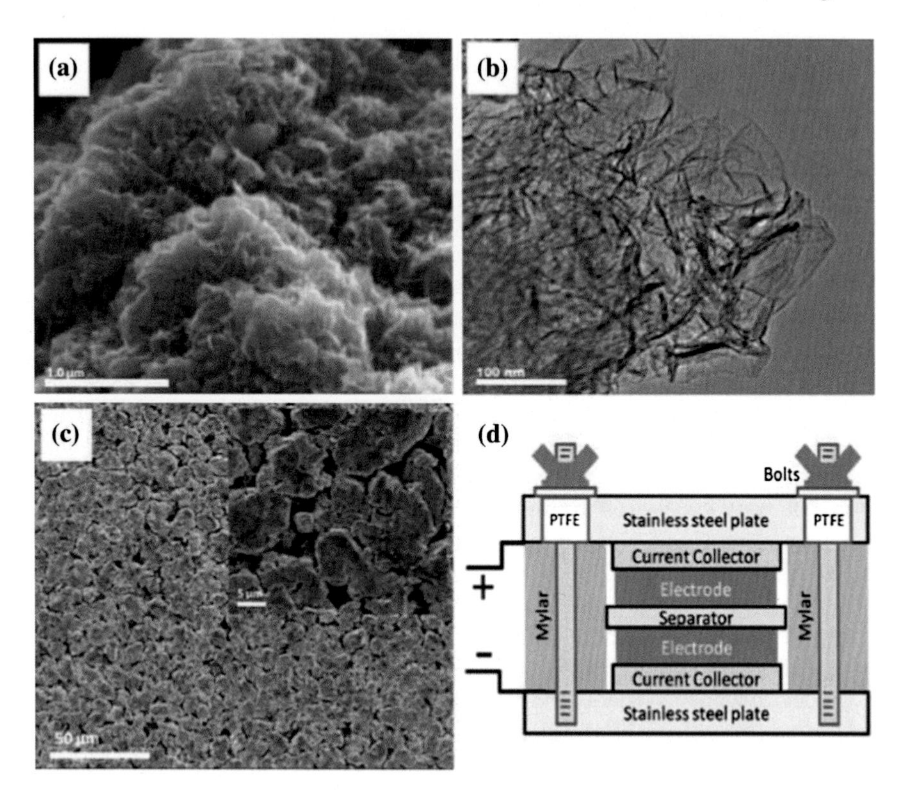

Fig. 1.21 **a** SEM image of chemically modified graphene (CMG) particle surface, **b** TEM image showing individual graphene sheets extending from CMG particle surface, **c** low and high (*inset*) magnification SEM images of CMG particle electrode surface, and **d** schematic of test cell assembly. Reprinted with the permission from Ref. [120]. Copyright 2008 American Chemical Society

and only the surface that wetted with solution is valid. It was confirmed that the capacitance increased from 14 F g^{-1} (the capacitance of dry and overlapped graphene) to 269 F g^{-1} by inserting Pt particles of 4 nm between the layers of graphene [122]. Obviously, the performance of graphene-based supercapacitor can be improved by constructing well-designed macrostructure. Carbon nanotubes are also used to separate graphene sheets, and the corresponding capacitance can be increased to 120 F g^{-1} [123].

To reduce the agglomeration and increase the electrochemical effective surface area, graphene sheets can be used to carry the active materials with large pseudocapacitacne. The composite of MnO_2 nanocrystal on graphene surface has large capacitance of 197 F g^{-1}, which is much higher than that of graphene oxide (10.9 F g^{-1}) and MnO_2 block (6.8 F g^{-1}) [124]. The composite of $Ni(OH)_2$ growing on graphene displays an extremely high capacitance of 1335 F g^{-1} [125]. $Ni(OH)_2$ nanosheets grown on graphene sheets are obviously superior to severely oxidized graphene

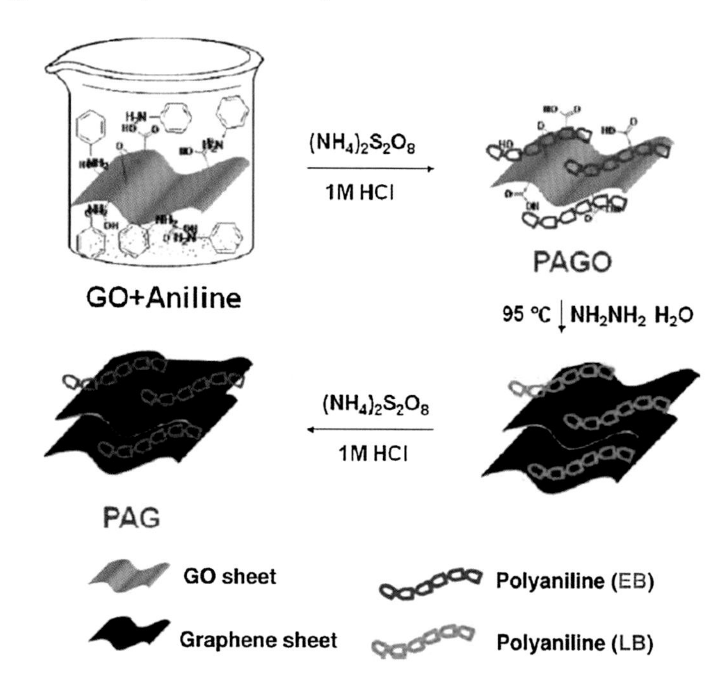

Fig. 1.22 Illustration of process for preparation of graphene/polyaniline composites for supercapacitors. *GO* graphene oxide, *PAGO* polyaniline/graphene oxide composite, *PAG* polyaniline/graphene composites, *EB* highly conductive half-oxidized emeraldine base of polyaniline, *LB* the reduced, neutral leucoemeraldine phase of polyaniline. Reprinted with the permission from Ref. [127]. Copyright 2010 American Chemical Society

oxide sheets, which nonconductive. This shows that the quality of bottom carrier and the structure design of nanoparticles are critical to the performance of supercapacitors. ZnO and SnO_2 modified graphene also has been reported, but the corresponding capacitance is unsatisfactory, which is around 62 F g^{-1} [126].

Graphene/polymer nanocomposites have a great potential in supercapacitor. Polyaniline/graphene-based nanomaterial can be easily prepared by electrochemical and chemical methods (Fig. 1.22). The capacitance ranged from 233 to 1046 F g^{-1} based on different composite nanostructure [127–131]. Poly (4-vinybenzene styrene sulfonic acide sodium) PPS-graphene nanocomposited capacitor has a very high cycle stability. After 14,860 cycles, its specific capacitance (190 F g^{-1}) only decreased by 12 % [132]. Graphene/polymer composite material is especially suitable for the portable and bendable electronic products, based on its toughness and unchanged performance under external pressure. Wu et al. displayed an excellent example that the capacitance of polyaniline/graphene composite material was up to 210 F g^{-1}, and the cycle stability was 94 % [133].

1.5.2.2 Li-Ion Cell

Rechargeable Li-ion cells (LIBs) are widely used in portable electronic devices, such as mobile phones, laptops, digital cameras, and others, which are considered to be a good choice for energy efficient and environmentally friendly equipment and facilities [134]. LIBs use the typical bulk storage method, and its cathode transfers the chemical energy into electrical energy by the intercalation and deintercalation of lithium ion in 3D graphite (Fig. 1.23). The energy density and performance of LIBs are mainly dependent on the physical and chemical properties of electrode materials. Therefore, many researches attempt to design novel nanostructures and develop new electrode materials [135, 136]. Graphene can be used as the carrier of pseudocapacitance active material (such as transition metal nanoparticles and electrical conductive polymer), forming composite nanomaterial, thus the energy density of system was greatly improved. Meanwhile, its 'anchoring' effect of surface defect can prevent agglomerate and shedding of active material, thereby extending the usage life of cell. Besides, graphene could replace the carbon nanotubes as cheap conductive filler, and be composited with $LiCoPO_4$ or other materials forming slurry to improve the anode conductivity and power density of LIB. However, the 2D network of graphene blocks the transfer tunnel of Li-ion, so the adding amount of graphene needs to be further adjusted so as to achieve the best result.

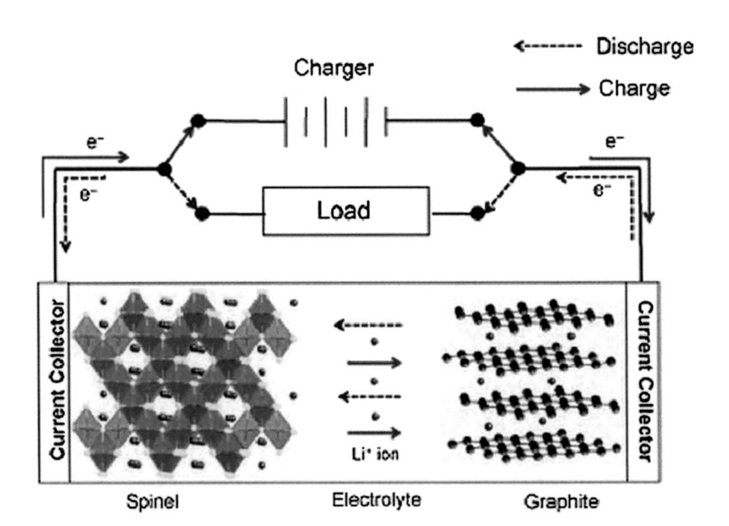

Fig. 1.23 Operating of LIB. The negative electrode is a graphitic carbon that holds Li in its layers, whereas the positive electrode is a Li-intercalation compound—usually an oxide because of its higher potential—that often is characterized by a layered structure. Both electrodes are able to reversibly insert and remove Li-ions from their respective structures. On charging, Li-ions are removed or deintercalated from the layered oxide compound and intercalated into the graphite layers. The process is reversed on discharge. The electrodes are separated by a nonaqueous electrolyte that transports Li-ions between the electrodes [137, 138]. Reprinted from Ref. [138], Copyright 2010, with permission from Elsevier

A series of graphene-based nanocarbons receive great attention as high capacitance electrode, including graphene nanosheets [139–141], graphene paper [142], and hybrid materials of graphene and carbon nanotubes or fullerene [143]. For a long time, researchers found that the specific capacitance of disordered carbon is higher than that of ordered carbon [144]. The intercalation mechanism of graphite/lithium ion (LiC_6) allows its ultimate capacitance to reach 372 mAh g^{-1}, while both sides of graphene, as completely exfoliated single-layer carbon crystal, can absorb lithium ions forming Li_3C, and the corresponding theoretical capacity is up to 744 mAh g^{-1}. So the overall capacitance of lithium ion cell is expected to be significantly improved [145]. Disordered assembly of graphene nanosheets can be prepared from graphene oxide through reduction, including hydrazine reduction, low temperature pyrolysis, rapid expansion, electron beam irradiation, etc. [139, 141, 146]. As the anode material of LIBs, these disordered graphenes generally exhibit higher lithium ion storage capacitance than that of graphite. This is mainly due to that amount of surface defects, large specific surface area, and more oxygen-containing functional groups are formed in chemically prepared graphene [16, 147, 148].

As mentioned above, graphene with large interlayer spacing is expected to have high specific capacitance. In fact, when we use CNTs or C_{60} as interlayer blocking agent to prevent graphene sheets to stack again, the overall capacitance of the battery will be further enhanced, increasing from 540 mAh g^{-1} to around 730 and 784 mAh g^{-1} [143]. Besides CNTs and C_{60}, metal or metal oxide nanoparticles are also used to composite with graphene sheets, including SnO_2 [149–154], TiO_2 [150, 155], Co_3O_4 [156–158], Mn_3O_4 [159], NiO [150], Fe_3O_4 [160, 161], CuO [162], $Co(OH)_2$ [163], Sn [164], Si [165], and $LiFePO_4$ [166]. These inorganic nanoparticles serve as not only the layer blocking agent between graphene layers, but also the active materials react reversibly with lithium ions. For example, as the anode of LIBs, SnO_2 has high theoretical specific capacity (782 mAh g^{-1}), while its cycle performance turns to bad due to the powdered process [167, 168]. When SnO_2 is composited with graphene sheets forming flexible 3D intercalated structure (Fig. 1.24), both the cycle performance and lithium storage capacitance would be significantly improved [149]. In this structure, the confinement effect of graphene on SnO_2 nanoparticles could greatly reduce the volume expansion caused by lithium intercalation, and the holes, formed between SnO_2 and graphene, provide buffer space for charging/discharging. In addition, graphene in the hybrid materials not only serves as the storage electrode of lithium ions, but also provides conductive tunnel to increase the electrochemical performance. It seems to be easy to composite graphene with inorganic nanoparticle, but in practice it is difficult, especially considering the controllability. Recently, Yang et al. reported a chemical method for large-scale preparation of 2D graphene-based mesoporous silicon wafer (GM-silica) with sandwich structure, wherein each graphene sheets are completely separated by mesoporous silica shell [157]. The obtained GM-Silica in sandwich structure can be used as template to be further assembled into graphene-based mesoporous carbon (GM-C) and Co_3O_4 (GM-Co_3O_4), using sucrose as carbon source, and cobalt nitride as cobalt source, respectively. By evaluating the lithium ions

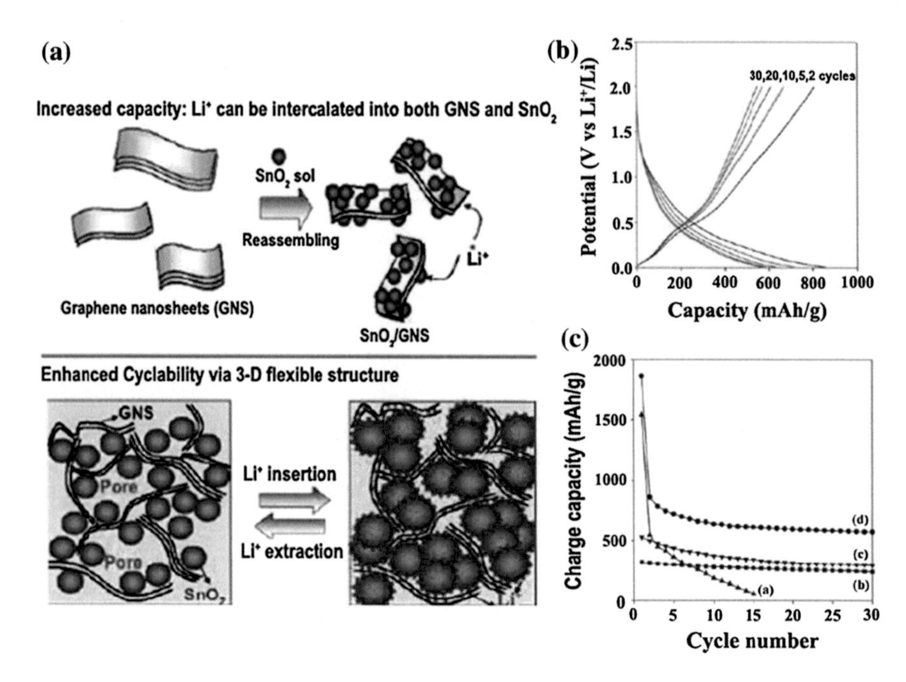

Fig. 1.24 Enhanced cyclic performance and lithium storage capacity of SnO_2/graphene nanoporous electrodes with three-dimensionally delaminated flexible structure. **a** Schematic illustration for the synthesis and the structure of SnO_2/graphene nanosheets (GNS). **b** Charge/discharge profile for SnO_2/GNS. **c** Cyclic performances for **a** bare SnO_2 nanoparticle, **b** graphite, **c** GNS, and **d** SnO_2/GNS. Reprinted with the permission from Ref. [149]. Copyright 2009 American Chemical Society

storage performance of GM-C sheet, it is found that the first cycle irreversible capacitance (915 mAh g^{-1}) is very high under magnification C/5. After 30 times cycle, both the discharge capacitance and charge capacitance of GM-C can be stabilized at 770 mAh g^{-1}, corresponding to the capacitance retention rate of 84 %. The result clearly shows that, the usage of graphene sheets could significantly enhance the electrochemical performance of LIBs by designing rational configuration [157].

In the above-mentioned nanomaterial, graphite oxide is the precursor of graphene, synthesized by top-down method. Bottom-up combined with in situ preparation is another way to prepare graphene-packaged metal or metal oxide nanoparticles, and the forming of particles and the packaging process occurred simultaneously in this method [170–172]. For example, it has been reported that carbon/metal or carbon/metal oxide packaging materials can be prepared by one-step pyrolysis of well-designed organic metal precursor [172]. The hollow Sn nanoparticles, with average diameter lower than 20 nm, are covered with 2–3 nm disordered carbon layer. The Sn content in this material is about 43 wt%. The stability of this material is good, as the reversible capacitance is as high as 550 mAh g^{-1} after 20 cycles, which is due to that the expansion and contraction of Sn particles are effectively relieved by carbon capsule. Carbon/Co_3O_4 composite also can be prepared by pyrolysis of organic Co precursor [171]. The first reversible capacitance of obtained graphene-coated Co_3O_4 is as high as 920 mAh g^{-1}, which is maintained at 940 mAh g^{-1} after 20 cycles. Compared with the bared Co_3O_4 nanoparticles, the great improvement of cycle performance is supposed to be related with the packing of particles by graphene and the defects in the graphene structure.

Generally speaking, the rate performance of LIBs was mainly controlled by the diffusion rate of lithium ions and the transport of electrons in electrode material. At present, many attempts have been tried to improve the rate performance of anode material, including increasing the diffusion coefficient, and decreasing the diffusion distance. One of the effective ways is to assemble graphene into 1D nanofiber, and directionally orient the graphene sheets vertical to the fiber axis, so that the diffusion distance of lithium ions is only half of the diameter of nanofibers [173]. This can be realized by CNT loading with aligned graphite vertical to axial direction [174]. However, it is still a big challenge for large-scale production of such 1D nanocarbon.

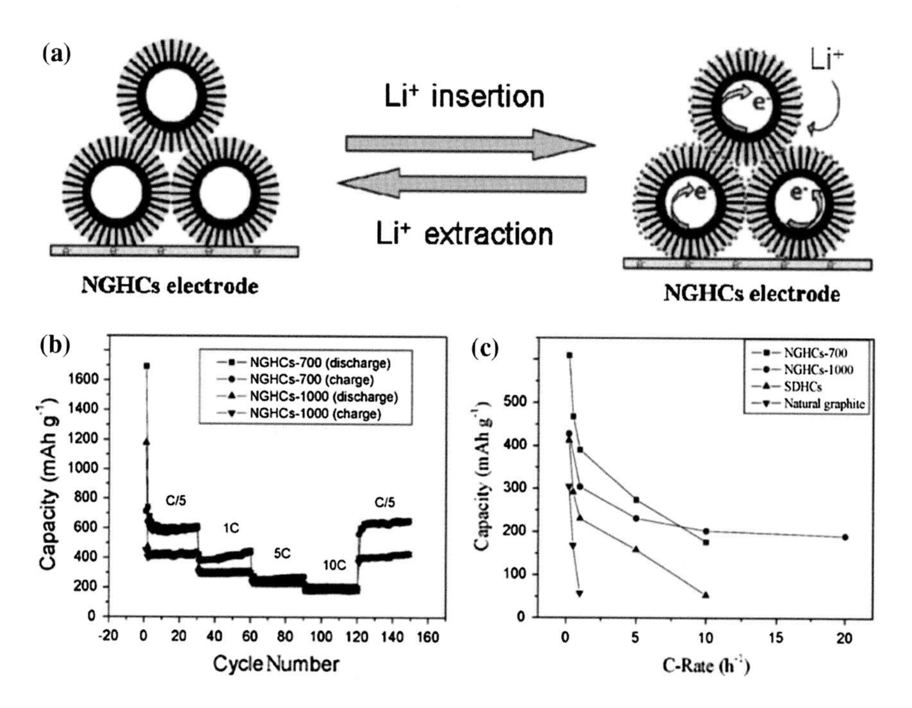

Fig. 1.25 **a** Schematic illustration of the diffusion of lithium ions and electrons during the discharge (*insertion*) and charge (*extraction*) processes of the nanographene-constructed hollow carbon spheres (NGHCs) electrode; **b** Rate performances of NGHC electrodes obtained at 700 °C (NGHCs-700) and 1000 °C (NGHCs-1000) at the rates of C/5, 1 C, 5 C, and 10 C); **c** Comparison of the rate capabilities of NGHCs-700, NGHCs-1000, SDHCs, and commercial natural-graphite electrodes. Reproduced from Ref. [169] by permission of John Wiley & Sons Ltd.

Recently, researcher designed and synthesized new nanographene constructed hollow carbon spheres (NGHCs) by controlled precursor pyrolysis method [169]. The nanoporous channel, constructed by graphene, is connected with the outer shell in the direction of vertical to spherical surface, which is in favor of Li-ion diffusion, and the carbon shell benefits the gather and diffusion of electrons at the same time. This special microstructure of NGHCs provides beneficially transmission mechanism for Li-ions and electrons (Fig. 1.25). So NGHCs showed high reversible capacitance (about 600 mAh g^{-1}) and rate performance (about 200 mAh g^{-1} at 10 °C) when used as the anode material of LIB (Fig. 1.25b) [169].

1.5.3 Other Applications

The unique 2D structure of graphene makes it widely apply in many fields, including advanced nanocomposites, gas detection of single molecule, integrated circuits, thermal interface materials, environment-friendly adsorbent materials, biodevices, antibacterial materials, etc. Besides, graphene can be applied as new interconnected material in integrated circuits by cutting it into specially designed nanobelt.

1.6 The Proposal of the Topic and the Main Research Contents

To ensure the security of energy strategy, each country invests heavily in the production, storage, and conversion of sustainable energy with low cost, high efficient, and green method. In our country, 'new energy' has been regarded as one of the seven new pillar industries by Twelfth Five-Year Plan, which promotes the quick expansion of new energy industry (such as wind power, photovoltaic industry, etc.) and significant increase of installed capacity. However, due to the instability and geographical constraints of new energy, how to efficiently store, release, and utilize at appropriate time and place is of particularly important. According to statistics, most wind power capacitance becomes 'garbage power' in our country due to inability of grid-connected operation (restricted by energy storage and other technology). Therefore, our country starts 'wind-light storage' engineering (Fig. 1.26), which is necessary in the initial stage of 'smart power grids', aiming at balancing the load of power grid by building large-scale energy storage facility, and realizing large-scale storage, and utilization of new energy. In addition, the development of advanced technology gives rise to new requirements for energy storage, including transportation facilities (electric or hybrid electric vehicles of energy conversation, emission reduction, and low carbon green), mobile electronics (communication equipment, laptops, cameras of lighter, thinner, and bendable), etc. The present energy storage system still has various problems, such as low energy density or

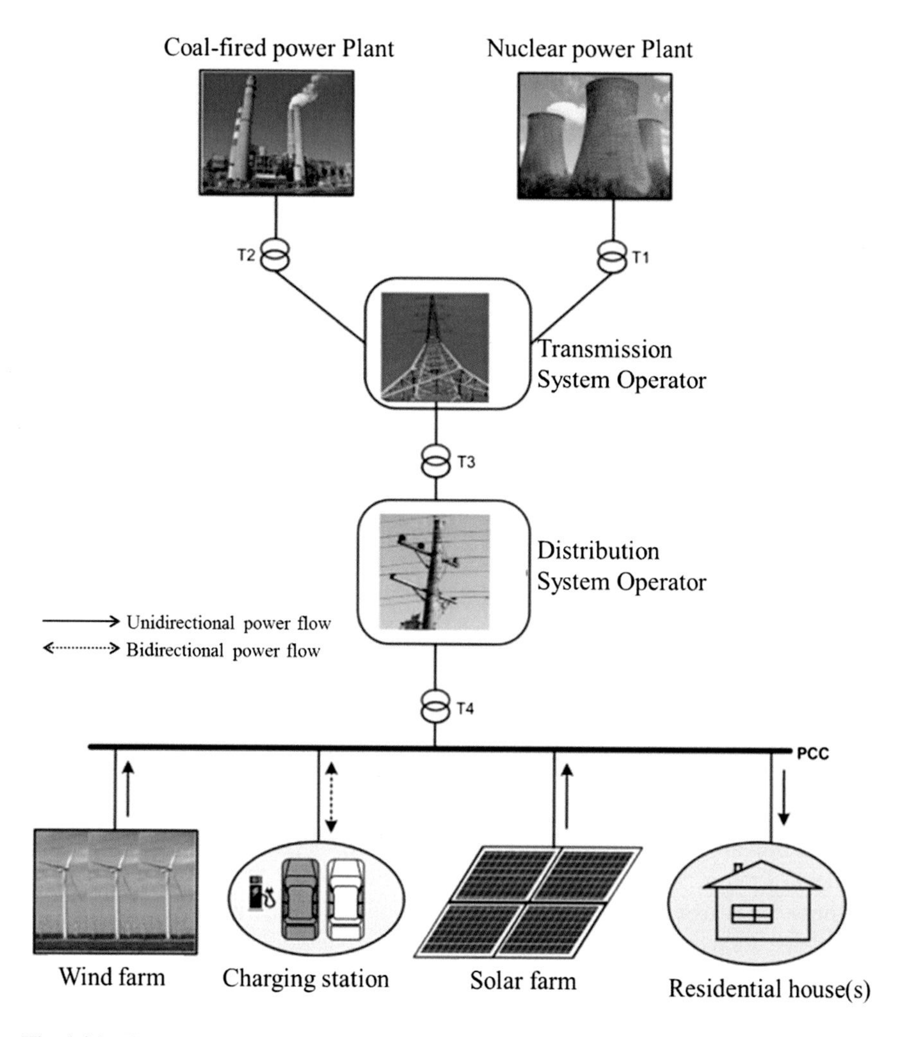

Fig. 1.26 The status of advanced energy conversion and storage for smart grid system. Reprinted from Ref. [175], Copyright 2014, with permission from Elsevier

power density, high cost, low energy efficient, environmental pollution brought by toxic material, etc., which bring opportunity and challenge for designing and developing new energy storage material.

Our country has rich graphite resource, producing and exporting the largest amount of graphite raw material in the world, but the further processing industry is backward relatively. Taking the opportunity of graphene research, we can develop new and practical further processing technology of graphite with independent intellectual property. The promotion and application of graphene in energy storage/conversion, environment protection, green catalyst, etc. can be realized by researching structure/functional combined graphene-based material. This work would accelerate the technical advance and industry upgrade of graphite processing industry. Research on graphene is expected to be another power source for scientific and technological progress.

Generally, there are abundant structural defects on the surfaces and edges of chemically derived graphene, such as functional groups (mainly –OH, –COOH, –C=O, –C–O–C–), lattice defects (atom vacancy, distortion, dangling bonds), etc. First, these defects are active sites of electrochemical energy storage, providing rich pseudocapacitance for supercapacitors and embedded sites for lithium ion. Second, the defects could be modified by nonmetal atoms (such as N, B, P, etc.) by further surface chemical reaction, thus the acid–base adjustment and electron modification are realized, which bring new surface/interface performance for electrochemical energy storage. At last, graphene can be used as new carrier and current collector of electrochemically active material (such as transition metal nanoparticles and conductive polymer) based on its typical sp^2 conjugated structure and anchoring effect of surface defect, forming graphene nanocomposite electrode. In this composite structure, the detachment and aggregation of active material can be avoided during charge/discharge cycle, corresponding to prolonged service time, while the energy and power density of the energy storage device are greatly increased at the same time. These special advantages provide tunable surface chemistry for the application of chemically derived graphene in advanced energy storage and conversion.

As a kind of nanomaterial, graphene could lower the volume capacity of the energy storage due to its extremely low density, but the employment of powder graphene increases the processing difficulty and cost, which becomes the major barrier for its practical application in energy storage. The surface wettability and solvent compatibility of chemically derived graphene are increased by introducing functional groups, which makes it possible for wet assembly of macrostructure. Graphene can be assembled into various macroscopic materials through 3D assembly technology, such as fiber, film, foam, bulk, fabric, felt, etc., while the nanoeffect is maintained at the same time. This type of method is expected to be an effective way to promote the large-scale application of graphene in energy storage field.

The work has been carried out following generally two mainlines including (a) tunable surface chemistry of graphene/graphene oxide and (b) graphene-based macroscopic assemblies/hybrids, with graphite as the starting material, toward application in advanced energy conversion and storage. The main results of present thesis are as follows:

(1) Chemically controllable preparation of graphene and its electrochemical performances: reduced graphene is prepared by vacuum-promoted thermal expansion of graphite oxide, combined with progressive carbonization to adjust the functionalization degree. This dissertation studied the evolution of micro texture and chemical structure during thermal annealing, and further built the bridge between structural evolution and electrochemical performance of TRGs.
(2) N-doped TRG and its electrochemical performances: the functional graphene prepared by vacuum-promoted low temperature expansion is progressively

aminated, obtaining hierarchically aminated graphene honeycombs. This dissertation studied the ratio and pattern of N-containing species at different temperatures, and proposed the chemical mechanism of N-doping. The energy storage properties, used in the electrode of supercapacitor, were also evaluated by analyzing the contribution mechanism of surface N/O heteroatoms to the pseudocapacitance of graphene.

(3) The structure evolution of graphene oxide film (GOF), self-assembled at a liquid/air interface, during annealing in confined space: free-standing GOF is obtained at liquid/air interface by original self-assembly technique, and then it is annealed in confined space to prepare flexible conductive graphene film. The chemical evolution mechanism of the obtained film is proposed by characterizing the evolution of microstructure and macroperformance during annealing.

(4) Assembly of 3D macroporous bubble graphene film and its capacitance: macroporous bubble graphene film with adjustable pore structure is built by directional and controllable assembly technique, among which polymer sphere is used as hard template. Then the film is assembled into binder-free supercapactior electrode to evaluate its energy storage performance. This dissertation studied the sandwich assembly mechanism between the template and graphene oxide, as well as the thermochemical mechanism of the reduced graphene oxide and the simultaneous removal of template.

(5) Selecting the synthesis way of Graphene/SnO_2 nanocomposite: graphene-supported SnO_2 hybrid material is prepared by two different methods, pre-(reduction followed by composite) and post-graphenized (composite followed by reduction) treatments. This dissertation also compared the differences in microstructure and macroscopic energy storage performance (supercapacitor and LIBs), and explained the intrinsic chemical and kinetic mechanisms of these two materials.

References

1. Geim AK, Novoselov KS. The rise of graphene. Nat Mater. 2007;6:183–91.
2. Eda G, Chhowalla M. Chemically derived graphene oxide: towards large-area thin-film electronics and optoelectronics. Adv Mater. 2010;22(22):2392–415.
3. Wallace PR. Erratum: the band theory of graphite. Phys Rev. 1947;72(3):258.
4. Novoselov KS, Geim AK, Morozov SV, Jiang D, Zhang Y, Dubonos SV, et al. Electric field effect in atomically thin carbon films. Science. 2004;306(5696):666–9.
5. Geim AK. Random walk to graphene (Nobel Lecture). Angew Chem Int Ed. 2011;50(31):6966–85.
6. Lee C, Wei X, Kysar JW, Hone J. Measurement of the elastic properties and intrinsic strength of monolayer graphene. Science. 2008;321(5887):385–8.
7. Zhu Y, Murali S, Cai W, Li X, Suk JW, Potts JR, et al. Graphene and graphene oxide: synthesis, properties, and applications. Adv Mater. 2010;22(35):3906–24.
8. Ran JL (2008) Graphene transistor with high speed of 26 GHz is reported by IBM. eNet Silicon Valley 2008. Cited; Available from: http://www.enet.com.cn/article/2008/1223/A20081223407223.shtml.

9. Hu YH, Wang H, Hu B. Thinnest two-dimensional nanomaterial-graphene for solar energy. ChemSusChem. 2010;3(7):782–96.
10. Luo B, Liu S, Zhi L. Chemical approaches toward graphene-based nanomaterials and their applications in energy-related areas. Small. 2011;8(5):630–46.
11. Soldano C, Mahmood A, Dujardin E. Production, properties and potential of graphene. Carbon. 2010;48(8):2127–50.
12. Bai H, Li C, Shi G. Functional composite materials based on chemically converted graphene. Adv Mater. 2011;23(9):1089–115.
13. Compton OC, Nguyen ST. Graphene oxide, highly reduced graphene oxide, and graphene: versatile building blocks for carbon-based materials. Small. 2010;6(6):711–23.
14. Koch KR, Krause PF. Oxidation by Mn_2O_7-an impressive demonstration of the powerful oxidizing property of dimanganeseheptoxide. J Chem Educ. 1982;59(11):973–4.
15. Wissler M. Graphite and carbon powders for electrochemical applications. J Power Sour. 2006;156(2):142–50.
16. Szabo T, Berkesi O, Forgo P, Josepovits K, Sanakis Y, Petridis D, et al. Evolution of surface functional groups in a series of progressively oxidized graphite oxides. Chem Mater. 2006;18(11):2740–9.
17. Scholz W, Boehm HP. Graphite Oxide. 6. Structure of graphite oxide. Zeitschrift Fur Anorganische Und Allgemeine Chemie 1969; 369(3–6): 327.
18. Nakajima T, Mabuchi A, Hagiwara R. A new structure model of graphite oxide. Carbon. 1988;26(3):357–61.
19. Lerf A, He HY, Forster M, Klinowski J. Structure of graphite oxide revisited. J Phys Chem B. 1998;102(23):4477–82.
20. He HY, Riedl T, Lerf A, Klinowski J. Solid-state NMR studies of the structure of graphite oxide. J Phys Chem. 1996;100(51):19954–8.
21. Lerf A, He HY, Riedl T, Forster M, Klinowski J. C-13 and H-1 MAS NMR studies of graphite oxide and its chemically modified derivatives. Solid State Ionics. 1997;101:857–62.
22. Buchsteiner A, Lerf A, Pieper J. Water dynamics in graphite oxide investigated with neutron scattering. J Physl Chem B. 2006;110(45):22328–38.
23. Lerf A, Buchsteiner A, Pieper J, Schottl S, Dekany I, Szabo T, et al. Hydration behavior and dynamics of water molecules in graphite oxide. J Phys Chem Solids. 2006;67(5–6):1106–10.
24. Boehm HP, Scholz W. Der Verpuffungspunkt Des Graphitoxids. Zeitschrift Fur Anorganische Und Allgemeine Chemie. 1965; 335(1–2): 74.
25. Rodriguez AM, Jimenez PSV. Some new aspects of graphite oxidation at O-degrees-C in a liquid-medium—a mechanism proposal for oxidation to graphite oxide. Carbon. 1986;24(2): 163–7.
26. Hadzi D, Novak A. Infra-red spectra of graphitic oxide. T Faraday Soc. 1955;51(12):1614–20.
27. Daniel RD, Sungjin P, Christopher WB, Rodney SR. The chemistry of graphene oxide. Chem Soc Rev. 2010;39:228–40.
28. He HY, Klinowski J, Forster M, Lerf A. A new structural model for graphite oxide. Chem Phys Lett. 1998;287(1–2):53–6.
29. Hofmann C, Frenzel A, Csalan E. The constitution of graphite acid and its reactions. Liebigs Ann Chem. 1934;510:1–41.
30. Cai WW, Piner RD, Stadermann FJ, Park S, Shaibat MA, Ishii Y, et al. Synthesis and solid-state NMR structural characterization of (13)C-labeled graphite oxide. Science. 2008;321(5897):1815–7.
31. Gao W, Alemany LB, Ci LJ, Ajayan PM. New insights into the structure and reduction of graphite oxide. Nat Chem. 2009;1(5):403–8.
32. Szabo T, Berkesi O, Dekany I. DRIFT study of deuterium-exchanged graphite oxide. Carbon. 2005;43(15):3186–9.
33. Szabo T, Tombacz E, Illes E, Dekany I. Enhanced acidity and pH-dependent surface charge characterization of successively oxidized graphite oxides. Carbon. 2006;44(3):537–45.

34. Boukhvalov DW, Katsnelson MI. Modeling of graphite oxide. J Am Chem Soc. 2008;130(32): 10697–701.
35. Pei S, Cheng H-M. The reduction of graphene oxide. Carbon. 2012;50(9):3210–28.
36. Paredes JI, Villar-Rodil S, Martinez-Alonso A, Tascon JMD. Graphene oxide dispersions in organic solvents. Langmuir. 2008;24(19):10560–4.
37. Becerril HA, Mao J, Liu Z, Stoltenberg RM, Bao Z, Chen Y. Evaluation of solution-processed reduced graphene oxide films as transparent conductors. ACS Nano. 2008;2(3):463–70.
38. Gomez-Navarro C, Weitz RT, Bittner AM, Scolari M, Mews A, Burghard M, et al. Electronic transport properties of individual chemically reduced graphene oxide sheets. Nano Lett. 2007;7(11):3499–503.
39. Stankovich S, Dikin DA, Dommett GHB, Kohlhaas KM, Zimney EJ, Stach EA, et al. Graphene-based composite materials. Nature. 2006;442(7100):282–6.
40. Li ZY, Zhang WH, Luo Y, Yang JL, Hou JG. How graphene is cut upon oxidation? J Am Chem Soc 2009; 131(18): 6320.
41. Zhang L, Liang JJ, Huang Y, Ma YF, Wang Y, Chen YS. Size-controlled synthesis of graphene oxide sheets on a large scale using chemical exfoliation. Carbon. 2009;47(14):3365–8.
42. Si Y, Samulski ET. Synthesis of water soluble graphene. Nano Lett. 2008;8(6):1679–82.
43. Wu ZS, Ren WC, Gao LB, Liu BL, Jiang CB, Cheng HM. Synthesis of high-quality graphene with a pre-determined number of layers. Carbon. 2009;47(2):493–9.
44. Schniepp HC, Li JL, McAllister MJ, Sai H, Herrera-Alonso M, Adamson DH, et al. Functionalized single graphene sheets derived from splitting graphite oxide. J Phys Chem B. 2006;110(17):8535–9.
45. McAllister MJ, Li J-L, Adamson DH, Schniepp HC, Abdala AA, Liu J, et al. Single sheet functionalized graphene by oxidation and thermal expansion of graphite. Chem Mater. 2007;19(18):4396–404.
46. Wu ZS, Ren WC, Gao LB, Zhao JP, Chen ZP, Liu BL, et al. Synthesis of graphene sheets with high electrical conductivity and good thermal stability by hydrogen arc discharge exfoliation. ACS Nano. 2009;3(2):411–7.
47. Kudin KN, Ozbas B, Schniepp HC, Prud'homme RK, Aksay IA, Car R. Raman spectra of graphite oxide and functionalized graphene sheets. Nano Lett. 2008;8(1):36–41.
48. Wang X, Zhi LJ, Mullen K. Transparent, conductive graphene electrodes for dye-sensitized solar cells. Nano Lett. 2008;8(1):323–7.
49. Zhao JP, Pei SF, Ren WC, Gao LB, Cheng HM. Efficient preparation of large-area graphene oxide sheets for transparent conductive films. ACS Nano. 2010;4(9):5245–52.
50. Mattevi C, Eda G, Agnoli S, Miller S, Mkhoyan KA, Celik O, et al. Evolution of electrical, chemical, and structural properties of transparent and conducting chemically derived graphene thin films. Adv Funct Mater. 2009;19(16):2577–83.
51. Yang D, Velamakanni A, Bozoklu G, Park S, Stoller M, Piner RD, et al. Chemical analysis of graphene oxide films after heat and chemical treatments by X-ray photoelectron and Micro-Raman spectroscopy. Carbon. 2009;47(1):145–52.
52. Pan DY, Zhang JC, Li Z, Wu MH. Hydrothermal route for cutting graphene sheets into blue-luminescent graphene quantum dots. Adv Mater 2010; 22(6): 734.
53. Li XL, Wang HL, Robinson JT, Sanchez H, Diankov G, Dai HJ. Simultaneous nitrogen doping and reduction of graphene oxide. J Am Chem Soc. 2009;131(43):15939–44.
54. Gengler RYN, Veligura A, Enotiadis A, Diamanti EK, Gournis D, Józsa C, et al. Large-yield preparation of high-electronic-quality graphene by a Langmuir-Schaefer approach. Small. 2010;6(1):35–9.
55. Su Q, Pang SP, Alijani V, Li C, Feng XL, Mullen K. Composites of graphene with large aromatic molecules. Adv Mater 2009; 21(31): 3191.
56. Brunauer S, Emmett PH, Teller E. Adsorption of gases in multimolecular layers. J Am Chem Soc. 1938;60:309–19.
57. Das A, Chakraborty B, Sood AK. Raman spectroscopy of graphene on different substrates and influence of defects. B Mater Sci. 2008;31(3):579–84.
58. Geim AK, Novoselov KS. The rise of graphene. Nat Mater. 2007;6(3):183–91.

59. Warner JH, Rummeli MH, Ge L, Gemming T, Montanari B, Harrison NM, et al. Structural transformations in graphene studied with high spatial and temporal resolution. Nat Nanotechnol. 2009;4(8):500–4.
60. Stankovich S, Dikin DA, Piner RD, Kohlhaas KA, Kleinhammes A, Jia Y, et al. Synthesis of graphene-based nanosheets via chemical reduction of exfoliated graphite oxide. Carbon. 2007;45(7):1558–65.
61. Kotov NA, Dekany I, Fendler JH. Ultrathin graphite oxide-polyelectrolyte composites prepared by self-assembly: transition between conductive and non-conductive states. Adv Mater 1996; 8(8): 637.
62. Stankovich S, Piner RD, Chen XQ, Wu NQ, Nguyen ST, Ruoff RS. Stable aqueous dispersions of graphitic nanoplatelets via the reduction of exfoliated graphite oxide in the presence of poly (sodium 4-styrenesulfonate). J Mater Chem. 2006;16(2):155–8.
63. Li D, Muller MB, Gilje S, Kaner RB, Wallace GG. Processable aqueous dispersions of graphene nanosheets. Nat Nanotechnol. 2008;3(2):101–5.
64. Fernandez-Merino MJ, Guardia L, Paredes JI, Villar-Rodil S, Solis-Fernandez P, Martinez-Alonso A, et al. Vitamin C is an ideal substitute for hydrazine in the reduction of graphene oxide suspensions. J Phys Chem C. 2010;114(14):6426–32.
65. Zhu YW, Cai WW, Piner RD, Velamakanni A, Ruoff RS. Transparent self-assembled films of reduced graphene oxide platelets. Appl Phys Lett. 2009;95(10):103104–7.
66. Robinson JT, Zalalutdinov M, Baldwin JW, Snow ES, Wei ZQ, Sheehan P, et al. Wafer-scale reduced graphene oxide films for nanomechanical devices. Nano Lett. 2008;8(10):3441–5.
67. Gilje S, Han S, Wang M, Wang KL, Kaner RB. A chemical route to graphene for device applications. Nano Lett. 2007;7(11):3394–8.
68. Chen H, Muller MB, Gilmore KJ, Wallace GG, Li D. Mechanically strong, electrically conductive, and biocompatible graphene paper. Adv Mater 2008; 20(18): 3557.
69. He QY, Sudibya HG, Yin ZY, Wu SX, Li H, Boey F, et al. Centimeter-long and large-scale micropatterns of reduced graphene oxide films: fabrication and sensing applications. ACS Nano; 4(6): 3201–8.
70. Zhou XZ, Huang X, Qi XY, Wu SX, Xue C, Boey FYC, et al. In situ synthesis of metal nanoparticles on single-layer graphene oxide and reduced graphene oxide surfaces. J Phys Chem C. 2009;113(25):10842–6.
71. Qi XY, Pu KY, Li H, Zhou XZ, Wu SX, Fan QL, et al. Amphiphilic graphene composites. Angew Chem Int Ed. 2010;49(49):9426–9.
72. Yin ZY, Wu SX, Zhou XZ, Huang X, Zhang QC, Boey F, et al. Electrochemical deposition of ZnO nanorods on transparent reduced graphene oxide electrodes for hybrid solar cells. Small. 2010;6(2):307–12.
73. Qi XY, Pu KY, Zhou XZ, Li H, Liu B, Boey F, et al. Conjugated-polyelectrolyte-functionalized reduced graphene oxide with excellent solubility and stability in polar solvents. Small. 2010;6(5):663–9.
74. He QY, Wu SX, Gao S, Cao XH, Yin ZY, Li H, et al. Transparent, flexible, all-reduced graphene oxide thin film transistors. ACS Nano. 2011;5(6):5038–44.
75. Shin H-J, Kim KK, Benayad A, Yoon S-M, Park HK, Jung I-S, et al. Efficient reduction of graphite oxide by sodium borohydride and its effect on electrical conductance. Adv Funct Mater. 2009;19(12):1987–92.
76. Periasamy M, Thirumalaikumar P. Methods of enhancement of reactivity and selectivity of sodium borohydride for applications in organic synthesis. J Organomet Chem. 2000;609(1–2):137–51.
77. Pei SF, Zhao JP, Du JH, Ren WC, Cheng HM. Direct reduction of graphene oxide films into highly conductive and flexible graphene films by hydrohalic acids. Carbon. 2010;48(15): 4466–74.
78. Moon IK, Lee J, Ruoff RS, Lee H. Reduced graphene oxide by chemical graphitization. Nat Commun. 2010;1:73.
79. Wang GX, Yang J, Park J, Gou XL, Wang B, Liu H, et al. Facile synthesis and characterization of graphene nanosheets. J Phys Chem C. 2008;112(22):8192–5.

80. Fan XB, Peng WC, Li Y, Li XY, Wang SL, Zhang GL, et al. Deoxygenation of exfoliated graphite oxide under alkaline conditions: a green route to graphene preparation. Adv Mater. 2008;20(23):4490–3.
81. Zhou XJ, Zhang JL, Wu HX, Yang HJ, Zhang JY, Guo SW. Reducing graphene oxide via hydroxylamine: a simple and efficient route to graphene. J Phys Chem C. 2011;115(24): 11957–61.
82. Ballard DGH, Rideal GR. Flexible inorganic films and coatings. J Mater Sci. 1983;18(2): 545–61.
83. Dikin DA, Stankovich S, Zimney EJ, Piner RD, Dommett GHB, Evmenenko G, et al. Preparation and characterization of graphene oxide paper. Nature. 2007;448(7152):457–60.
84. Williams G, Seger B, Kamat PV. TiO_2-graphene nanocomposites. UV-assisted photocatalytic reduction of graphene oxide. ACS Nano. 2008;2(7):1478–91.
85. Ian VL, Thomas HK, Prashant VK. Anchoring semiconductor and metal nanoparticles on a two-dimentional catalyst mat. Storing and shuttling electrons with reduced graphene oxide. Nano Lett. 2010;10:577–83.
86. Georgios KD, Emmanuel T, George EF. Pillared graphene: a new 3-D network nanostructure for enhanced hydrogen storage. Nano Lett. 2008;8(10):3166–70.
87. Bourlinos AB, Gournis D, Petridis D, Szabo T, Szeri A, Dekany I. Graphite oxide: chemical reduction to graphite and surface modification with primary aliphatic amines and amino acids. Langmuir. 2003;19(15):6050–5.
88. Eda G, Fanchini G, Chhowalla M. Large-area ultrathin films of reduced graphene oxide as a transparent and flexible electronic material. Nat Nanotechnol. 2008;3(5):270–4.
89. Zhang JL, Yang HJ, Shen GX, Cheng P, Zhang JY, Guo SW. Reduction of graphene oxide via L-ascorbic acid. Chem Commun. 2010;46(7):1112–4.
90. Cote LJ, Kim F, Huang JX. Langmuir-blodgett assembly of graphite oxide single layers. J Am Chem Soc. 2009;131(3):1043–9.
91. Brunetaud X, Divet L, Damidot D. Impact of unrestrained Delayed Ettringite Formation-induced expansion on concrete mechanical properties. Cement Concrete Res. 2008;38(11): 1343–8.
92. Compton OC, Dikin DA, Putz KW, Brinson LC, Nguyen ST. Electrically conductive "alkylated" graphene paper via chemical reduction of amine-functionalized graphene oxide paper. Adv Mater 2010; 22(8): 892.
93. Park S, Lee KS, Bozoklu G, Cai W, Nguyen ST, Ruoff RS. Graphene oxide papers modified by divalent ions—Enhancing mechanical properties via chemical cross-linking. ACS Nano. 2008;2(3):572–8.
94. Berhan L, Yi YB, Sastry AM, Munoz E, Selvidge M, Baughman R. Mechanical properties of nanotube sheets: alterations in joint morphology and achievable moduli in manufacturable materials. J Appl Phys. 2004;95(8):4335–45.
95. Seah KHW, Hemanth J, Sharma SC. Tensile strength and hardness of sub-zero chilled cast iron. Mater Design. 1995;16(3):175–9.
96. Chen CM, Yang Q-H, Yang Y, Lv W, Wen Y, Hou P-X, et al. Self-assembled free-standing graphite oxide membrane. Adv Mater. 2009;21(29):3007–11.
97. Li D, Kaner RB. Materials science-graphene-based materials. Science. 2008;320(5880): 1170–1.
98. Liu ZH, Wang ZM, Yang XJ, Ooi KT. Intercalation of organic ammonium ions into layered graphite oxide. Langmuir. 2002;18(12):4926–32.
99. Li XL, Zhang GY, Bai XD, Sun XM, Wang XR, Wang E, et al. Highly conducting graphene sheets and Langmuir-Blodgett films. Nat Nanotechnol. 2008;3(9):538–42.
100. Park S, An JH, Jung IW, Piner RD, An SJ, Li XS, et al. Colloidal suspensions of highly reduced graphene oxide in a wide variety of organic solvents. Nano Lett. 2009;9(4):1593–7.
101. Park S, An JH, Piner RD, Jung I, Yang DX, Velamakanni A, et al. Aqueous suspension and characterization of chemically modified graphene sheets. Chem Mater. 2008;20(21):6592–4.
102. Xu YX, Bai H, Lu GW, Li C, Shi GQ. Flexible graphene films via the filtration of water-soluble noncovalent functionalized graphene sheets. J Am Chem Soc 2008; 130(18): 5856.

103. El-Kady MF, Strong V, Dubin S, Kaner RB. Laser scribing of high-performance and flexible graphene-based electrochemical capacitors. Science. 2012;335(6074):1326–30.
104. Obraztsov AN. Chemical vapour deposition making graphene on a large scale. Nat Nanotechnol. 2009;4(4):212–3.
105. Emtsev KV, Bostwick A, Horn K, Jobst J, Kellogg GL, Ley L, et al. Towards wafer-size graphene layers by atmospheric pressure graphitization of silicon carbide. Nat Mater. 2009;8(3):203–7.
106. Orlita M, Faugeras C, Plochocka P, Neugebauer P, Martinez G, Maude DK, et al. Approaching the dirac point in high-mobility multilayer epitaxial graphene. Phys Rev Lett 2008; 101(26).
107. Castro Neto AH, Guinea F, Peres NMR, Novoselov KS, Geim AK. The electronic properties of graphene. Rev Mod Phys 2009; 81(1): 109–62.
108. Wu JB, Becerril HA, Bao ZN, Liu ZF, Chen YS, Peumans P. Organic solar cells with solution-processed graphene transparent electrodes. Appl Phys Lett 2008; 92(26).
109. Frackowiak E. Carbon materials for supercapacitor application. Phys Chem Chem Phys. 2007;9(15):1774–85.
110. Zhang LL, Zhou R, Zhao XS. Graphene-based materials as supercapacitor electrodes. J Mater Chem. 2010;20(29):5983–92.
111. Conway BE, editor. Electrochemical supercapacitors: scientific fundamentals and technological applications. New York: Kluwer Academic/Plenum Publisher; 1999.
112. Hu CC, Chang KH, Lin MC, Wu YT. Design and tailoring of the nanotubular arrayed architecture of hydrous RuO_2 for next generation supercapacitors. Nano Lett. 2006;6(12): 2690–5.
113. Zhang LL, Wei TX, Wang WJ, Zhao XS. Manganese oxide-carbon composite as supercapacitor electrode materials. Micropor Mesopor Mat. 2009;123(1–3):260–7.
114. Zhang H, Cao GP, Wang ZY, Yang YS, Shi ZJ, Gu ZN. Growth of manganese oxide nanoflowers on vertically-aligned carbon nanotube arrays for high-rate electrochemical capacitive energy storage. Nano Lett. 2008;8(9):2664–8.
115. Choi D, Blomgren GE, Kumta PN. Fast and reversible surface redox reaction in nanocrystalline vanadium nitride supercapacitors. Adv Mater 2006; 18(9): 1178.
116. Fan LZ, Hu YS, Maier J, Adelhelm P, Smarsly B, Antonietti M. High electroactivity of polyaniline in supercapacitors by using a hierarchically porous carbon monolith as a support. Adv Funct Mater. 2007;17(16):3083–7.
117. Zhang LL, Li S, Zhang JT, Guo PZ, Zheng JT, Zhao XS. Enhancement of electrochemical performance of macroporous carbon by surface coating of polyaniline. Chem Mater. 2010;22(3):1195–202.
118. Seredych M, Hulicova-Jurcakova D, Lu GQ, Bandosz TJ. Surface functional groups of carbons and the effects of their chemical character, density and accessibility to ions on electrochemical performance. Carbon. 2008;46(11):1475–88.
119. Pumera M. Graphene-based nanomaterials for energy storage. Ener Environ Sci. 2011;4(3): 668–74.
120. Stoller MD, Park SJ, Zhu YW, An JH, Ruoff RS. Graphene-based ultracapacitors. Nano Lett. 2008;8(10):3498–502.
121. Wang Y, Shi ZQ, Huang Y, Ma YF, Wang CY, Chen MM, et al. Supercapacitor devices based on graphene materials. J Phys Chem C. 2009;113(30):13103–7.
122. Si YC, Samulski ET. Exfoliated graphene separated by platinum nanoparticles. Chem Mater. 2008;20(21):6792–7.
123. Yu DS, Dai LM. Self-assembled graphene/carbon nanotube hybrid films for supercapacitors. J Phys Chem Lett. 2010;1(2):467–70.
124. Chen S, Zhu JW, Wu XD, Han QF, Wang X. Graphene oxide-MnO_2 nanocomposites for supercapacitors. ACS Nano. 2010;4(5):2822–30.
125. Wang HL, Casalongue HS, Liang YY, Dai HJ. $Ni(OH)_2$ nanoplates grown on graphene as advanced electrochemical pseudocapacitor materials. J Am Chem Soc. 2010;132(21): 7472–7.

126. Lu T, Zhang YP, Li HB, Pan LK, Li YL, Sun Z. Electrochemical behaviors of graphene-ZnO and graphene-SnO_2 composite films for supercapacitors. Electrochim Acta. 2010;55(13): 4170–3.
127. Zhang K, Zhang LL, Zhao XS, Wu JS. Graphene/polyaniline nanofiber composites as supercapacitor electrodes. Chem Mater. 2010;22(4):1392–401.
128. Wang DW, Li F, Zhao JP, Ren WC, Chen ZG, Tan J, et al. Fabrication of graphene/polyaniline composite paper via in situ anodic electropolymerization for high-performance flexible electrode. ACS Nano. 2009;3(7):1745–52.
129. Murugan AV, Muraliganth T, Manthiram A. Rapid, facile microwave-solvothermal synthesis of graphene nanosheets and their polyaniline nanocomposites for energy strorage. Chem Mater. 2009;21(21):5004–6.
130. Yan J, Wei T, Shao B, Fan ZJ, Qian WZ, Zhang ML, et al. Preparation of a graphene nanosheet/polyaniline composite with high specific capacitance. Carbon. 2010;48(2):487–93.
131. Wang HL, Hao QL, Yang XJ, Lu LD, Wang X. Graphene oxide doped polyaniline for supercapacitors. Electrochem Commun. 2009;11(6):1158–61.
132. Jeong HK, Jin M, Ra EJ, Sheem KY, Han GH, Arepalli S, et al. Enhanced electric double layer capacitance of graphite oxide intercalated by poly(sodium 4-styrensulfonate) with high cycle stability. ACS Nano. 2010;4(2):1162–6.
133. Wu Q, Xu YX, Yao ZY, Liu AR, Shi GQ. Supercapacitors based on flexible graphene/polyaniline nanofiber composite films. ACS Nano. 2010;4(4):1963–70.
134. Tarascon JM, Armand M. Issues and challenges facing rechargeable lithium batteries. Nature. 2001;414(6861):359–67.
135. Liu C, Li F, Ma LP, Cheng HM. Advanced materials for energy storage. Adv Mater. 2010;22(8):E28–62.
136. Bruce PG, Scrosati B, Tarascon JM. Nanomaterials for rechargeable lithium batteries. Angew Chem Int Ed. 2008;47(16):2930–46.
137. Dunn B, Kamath H, Tarascon JM. Electrical energy storage for the grid: a battery of choices. Science. 2011;334(6058):928–35.
138. Park M, Zhang XC, Chung M, Less GB, Sastry AM. A review of conduction phenomena in Li-ion batteries. J Power Sources. 2010;195(24):7904–29.
139. Guo P, Song HH, Chen XH. Electrochemical performance of graphene nanosheets as anode material for lithium-ion batteries. Electrochem Commun. 2009;11(6):1320–4.
140. Pan DY, Wang S, Zhao B, Wu MH, Zhang HJ, Wang Y, et al. Li storage properties of disordered graphene nanosheets. Chem Mater. 2009;21(14):3136–42.
141. Wang GX, Shen XP, Yao J, Park J. Graphene nanosheets for enhanced lithium storage in lithium ion batteries. Carbon. 2009;47(8):2049–53.
142. Wang CY, Li D, Too CO, Wallace GG. Electrochemical properties of graphene paper electrodes used in lithium batteries. Chem Mater. 2009;21(13):2604–6.
143. Yoo E, Kim J, Hosono E, Zhou H, Kudo T, Honma I. Large reversible Li storage of graphene nanosheet families for use in rechargeable lithium ion batteries. Nano Lett. 2008;8(8):2277–82.
144. Sato K, Noguchi M, Demachi A, Oki N, Endo M. A mechanism of lithium storage in disordered carbons. Science. 1994;264(5158):556–8.
145. Dahn JR, Zheng T, Liu YH, Xue JS. Mechanisms for lithium insertion in carbonaceous materials. Science. 1995;270(5236):590–3.
146. Liang MH, Zhi LJ. Graphene-based electrode materials for rechargeable lithium batteries. J Mater Chem. 2009;19(33):5871–8.
147. Wang S, Chia PJ, Chua LL, Zhao LH, Png RQ, Sivaramakrishnan S, et al. Band-like transport in surface-functionalized highly solution-processable graphene nanosheets. Adv Mater 2008; 20(18): 3440.
148. Radovic LR, Bockrath B. On the chemical nature of graphene edges: Origin of stability and potential for magnetism in carbon materials. J Am Chem Soc. 2005;127(16):5917–27.

149. Paek SM, Yoo E, Honma I. Enhanced cyclic performance and lithium storage capacity of SnO_2/graphene nanoporous electrodes with three-dimensionally delaminated flexible structure. Nano Lett. 2009;9(1):72–5.
150. Wang DH, Kou R, Choi D, Yang ZG, Nie ZM, Li J, et al. Ternary self-assembly of ordered metal oxide-graphene nanocomposites for electrochemical energy storage. ACS Nano. 2010;4(3):1587–95.
151. Yao J, Shen XP, Wang B, Liu HK, Wang GX. In situ chemical synthesis of SnO_2-graphene nanocomposite as anode materials for lithium-ion batteries. Electrochem Commun. 2009;11(10):1849–52.
152. Zhang LS, Jiang LY, Yan HJ, Wang WD, Wang W, Song WG, et al. Mono dispersed SnO_2 nanoparticles on both sides of single layer graphene sheets as anode materials in Li-ion batteries. J Mater Chem. 2010;20(26):5462–7.
153. Li YM, Lv XJ, Lu J, Li JH. Preparation of SnO_2-nanocrystal/graphene-nanosheets composites and their lithium storage ability. J Phys Chem C. 2010;114(49):21770–4.
154. Du ZF, Yin XM, Zhang M, Hao QY, Wang YG, Wang TH. In situ synthesis of SnO_2/graphene nanocomposite and their application as anode material for lithium ion battery. Mater Lett. 2010;64(19):2076–9.
155. Yang SB, Feng XL, Mullen K. Sandwich-like, graphene-based titania nanosheets with high surface area for fast lithium storage. Adv Mater 2011; 23(31): 3575.
156. Yang SB, Cui GL, Pang SP, Cao Q, Kolb U, Feng XL, et al. Fabrication of cobalt and cobalt oxide/graphene composites: Towards high-performance anode materials for lithium ion batteries. ChemSusChem. 2010;3(2):236–9.
157. Yang SB, Feng XL, Ivanovici S, Mullen K. Fabrication of graphene-encapsulated oxide nanoparticles: Towards high-performance anode materials for lithium storage. Angew Chem Int Ed. 2010;49(45):8408–11.
158. Kim H, Seo DH, Kim SW, Kim J, Kang K. Highly reversible Co_3O_4/graphene hybrid anode for lithium rechargeable batteries. Carbon. 2011;49(1):326–32.
159. Wang HL, Cui LF, Yang YA, Casalongue HS, Robinson JT, Liang YY, et al. Mn_3O_4-graphene hybrid as a high-capacity anode material for lithium ion batteries. J Am Chem Soc. 2010;132(40):13978–80.
160. Li BJ, Cao HQ, Shao J, Qu MZ, Warner JH. Superparamagnetic Fe_3O_4 nanocrystals@graphene composites for energy storage devices. J Mater Chem. 2011;21(13):5069–75.
161. Zhou GM, Wang DW, Li F, Zhang LL, Li N, Wu ZS, et al. Graphene-wrapped Fe_3O_4 anode material with improved reversible capacity and cyclic stability for lithium ion batteries. Chem Mater. 2010;22(18):5306–13.
162. Mai YJ, Wang XL, Xiang JY, Qiao YQ, Zhang D, Gu CD, et al. CuO/graphene composite as anode materials for lithium-ion batteries. Electrochim Acta. 2011;56(5):2306–11.
163. He YS, Bai DW, Yang XW, Chen J, Liao XZ, Ma ZF. A $Co(OH)_2$-graphene nanosheets composite as a high performance anode material for rechargeable lithium batteries. Electrochem Commun. 2010;12(4):570–3.
164. Wang GX, Wang B, Wang XL, Park J, Dou SX, Ahn H, et al. Sn/graphene nanocomposite with 3D architecture for enhanced reversible lithium storage in lithium ion batteries. J Mater Chem. 2009;19(44):8378–84.
165. Lee JK, Smith KB, Hayner CM, Kung HH. Silicon nanoparticles-graphene paper composites for Li ion battery anodes. Chem Commun. 2010;46(12):2025–7.
166. Wang L, Wang HB, Liu ZH, Xiao C, Dong SM, Han PX, et al. A facile method of preparing mixed conducting $LiFePO_4$/graphene composites for lithium-ion batteries. Solid State Ionics. 2010;181(37–38):1685–9.
167. Huang JY, Zhong L, Wang CM, Sullivan JP, Xu W, Zhang LQ, et al. In situ observation of the electrochemical lithiation of a single SnO_2 nanowire electrode. Science. 2010;330(6010):1515–20.
168. Idota Y, Kubota T, Matsufuji A, Maekawa Y, Miyasaka T. Tin-based amorphous oxide: a high-capacity lithium-ion-storage material. Science. 1997;276(5317):1395–7.

169. Yang SB, Feng XL, Zhi LJ, Cao QA, Maier J, Mullen K. Nanographene-constructed hollow carbon spheres and their favorable electroactivity with respect to lithium storage. Adv Mater 2010; 22(7): 838.
170. Zhi LJ, Hu YS, El Hamaoui B, Wang X, Lieberwirth I, Kolb U, et al. Precursor-controlled formation of novel carbon/metal and carbon/metal oxide nanocomposites. Adv Mater 2008; 20(9): 1727.
171. Cui GL, Gu L, Zhi LJ, Kaskhedikar N, van Aken PA, Mullen K, et al. A germanium-carbon nanocomposite material for lithium batteries. Adv Mater. 2008;20(16):3079–83.
172. Cui GL, Hu YS, Zhi LJ, Wu DQ, Lieberwirth I, Maier J, et al. A one-step approach towards carbon-encapsulated hollow tin nanoparticles and their application in lithium batteries. Small. 2007;3(12):2066–9.
173. Subramanian V, Zhu HW, Wei BQ. High rate reversibility anode materials of lithium batteries from vapor-grown carbon nanofibers. J Phys Chem B. 2006;110(14):7178–83.
174. Zhi LJ, Wu JS, Li JX, Kolb U, Mullen K. Carbonization of disclike molecules in porous alumina membranes: Toward carbon nanotubes with controlled graphene-layer orientation. Angew Chem Int Ed. 2005;44(14):2120–3.
175. Mwasilu F, Justo JJ, Kim EK, Do TD, Jung JW. Electric vehicles and smart grid interaction: a review on vehicle to grid and renewable energy sources integration. Renew Sustain Energy Rev. 2014;34:501–16.

Chapter 2
Structural Evolution of the Thermally Reduced Graphene Nanosheets During Annealing

2.1 Introduction

Supercapacitor, also called electrochemical capacitor or ultracapacitor, is considered to be one of the newest innovations in the field of electrical energy storage. Compared to the battery devices, it is not limited by the electrochemical charge transfer kinetics of batteries, and thus owns unique advantages in high power density (10 kW kg^{-1}), short charge/discharge duration (in seconds), and long cycle life (over a million cycles) [1, 2]. These features have made supercapacitors very popular in various applications, such as hybrid electric vehicles, electric tools, and industrial power management [3].

Carbon materials are regarded as the first candidate electrode materials for supercapacitors [4, 5]. On one hand, charge storage on carbon electrodes is predominantly capacitive based on the electric double-layer capacitor (EDLC) through the electrostatic attraction of electrolyte ions onto the surface of carbon materials. On the other hand, contributions from surface functional groups can also be charged and discharged depending on Faradic redox reactions to yield pseudocapacitance (PC) [6].

Among various carbon allotropes, graphene shows competitive advantages in electrochemical energy storage applications, such as supercapacitors and Li ion batteries owing to its unique two-dimensional structure, high electronic conductivity, huge surface area, and good chemical stability [7–11]. Since the thermally reduced graphene (TRG) can be reliably obtained by thermal exfoliation of graphite oxide (GO), the chemical method became the strategic starting point for scale-up production of graphene [12–17]. As inherited from the precursor GO, TRG is instinctively decorated by various active sites (e.g., heteroatoms and lattice defects), which is essential with respect to providing a fertile ground for surface chemistry of carbon. The functionality, microtexture, polarization, acid/basic, as well as electronic properties of graphene can be easily tailored by adjusting the degree of thermal annealing [18]. The promising properties together

C.-M. Chen, *Surface Chemistry and Macroscopic Assembly of Graphene for Application in Energy Storage*, Springer Theses,
DOI 10.1007/978-3-662-48676-4_2

with the ease of processability and functionalization make graphene an ideal candidate for developing high performance supercapacitors [19–29]. It has reported that the ultimate performance of graphene-based supercapacitors will be bound up with the physical and chemical characteristics of graphene electrodes [22–24, 28]. Thus, a deep understanding of the structural evolution of graphene during annealing process is desirable for matching chemical surface properties with supercapacitor applications, which provides further new insights into the design of advanced energy storage devices in industry.

In this chapter, the TRG was produced by vacuum promoted thermal expansion of GO at relatively low temperatures, and the surface chemistry of which was further tuned by progressive annealing at various temperatures. Additionally, the structural evolution (microtexture, pore structure, type, and density of residue functionalities) of TRGs was investigated, while the electrochemical performance of which as supercapacitor electrodes was also evaluated.

2.2 Experimental

2.2.1 Preparation of TRG

The TRG was prepared by rapid heating up of GO under high vacuum [14]. GO was obtained by a modified Hummers' method [30]. The as-obtained GO was grounded into fine powder (~100 mesh), and air dried at 100 °C for 3.0 h. Then it was put into a quartz tube that was sealed at one end and stopped at the other end, through which the reactor was connected to a vacuum pump. The tube was heated at a rate of 30 °C min^{-1} in high vacuum (<2 Pa). At about 200 °C, an abrupt expansion was observed. The further annealing process of the above expanded GO was carried out at 250, 600, 800, and 1000 °C for 20 min, respectively. Finally, a series of graphene samples with different surface functionalities were obtained, which are named GT, where T represents the further annealing temperature. During the whole exfoliating and annealing process, a high vacuum in the quartz tube with a pressure of less than 5.0 Pa was maintained, as the vacuum pump was kept onto suck the desorbed gas out of the tube.

2.2.2 Sample Characterization

Scanning electron microscope (SEM) investigations were carried out with a JEOL JSM 7401F operated at 2.0 kV and a JEOL JEM 2010 transmission electron microscope (TEM) operated at 200.0 kV. The samples were ultrasonically dispersed in ethanol, and a drop of the solution was deposited on a Lacey carbon film grid for TEM characterization; X-ray diffraction (XRD) patterns were obtained at room temperature using specular reflection mode (Cu Kα radiation, λ = 0.15406 nm, D8 Advance, BRUKER/AXS, Germany); Laser Raman spectroscopy was performed on powder samples by using an ISA LabRam instrument equipped with an Olympus BX40 microscope. The excitation wavelength was 632.8 nm and a spectral resolution of 0.9 cm^{-1} was used; N_2 adsorption isotherm was measured using a Micromeritics 2375 at 77 K. The specific surface area was calculated by the Brunauer-Emmett-Teller (BET) method, the surface area of micropores (S_{micro}) was calculated by the t-plot approach (plot range from 3.5 to 5 nm), and the pore size distribution was deduced from the desorption isotherm by the Barret-Joyner-Halenda (BJH) method; Fourier transform infrared spectroscopy (FTIR) spectrums were carried out on a Thermo Nicolet IR200 spectrometer and the sample was pre-pressed with KBr into pellets before test; X-ray photoelectron spectroscopy (XPS) was obtained on the Thermo VG ESCALAB250 surface analysis system with parameters: Al Kα = 1486.6 eV, Power = 150 W (HV = 15 kV and I = 10 mA), spot size = 500 μm, pass energy 50.0 eV and energy step size 0.1 eV; the as-obtained graphene samples were further performed using thermal gravimetric analysis (TGA) system (STA 409 PC Luxx, Netzsch, Germany) equipped with a sweep gas mass spectroscopy appliance (ALZERS OmniStar 20, Switzerland).

2.2.3 *Electrochemical Measurements*

Electrochemical measurements of TRGs were performed using a three-electrode system in 6.0 M KOH aqueous solution, whereas Ni foam coated with electrode materials served as the working electrode, a platinum foil electrode as the counter electrode, and a reversible hydrogen electrode (RHE) as the reference electrode. The working electrodes were prepared as following: a mixture of the active material, carbon black, and poly (tetrafluoroethylene) (PTFE) with a weight ratio of 80:5:15 was ground together to form homogeneous slurry. The slurry was squeezed into a film and then punched into pellets (area ~0.8 cm^2). The punched pellets with a piece of nickel foam on each side were pressed under 2.5 MPa and dried overnight at 110 °C. Each working electrode contained ~3.5 mg electroactive materials. The electrodes were saturated with the electrolyte by vacuum enhanced impregnation for 2.0 h prior to the electrochemical evaluation. Cyclic voltammetry (CV) curves, galvanostatic charging/discharging (GC) curves and electrochemical impedance spectroscopy (EIS) profiles (frequency from 200 kHz to 10 mHz) were measured with a BioLogic electrochemistry workstation. The specific capacitances (C_F, F g^{-1}) were calculated from CV curves by equation: $C_F = S_{int}/(2\ V_s\ m)$, where S_{int} in mA V is the integrated area of CV curves, V_s in mV s^{-1} is the sweep rate, and m in g is the weight of active materials in the electrode. The Nyquist plots were fitted by the EC-Lab software with the equivalent circuit as $R_c + C_c/(R_i + W) + C_d$, the Randomize plus Levenberg–Marquardt method was employed for the fitting. The Ragone plot was calculated from GC results, the energy density (E, Wh kg^{-1}) was calculated by equation: $E = 1/2 * (C_F/4) * U^2$, where U is the potential window employed for GC (1.0 V in this work). While power density (P, W kg^{-1}) was calculated by: $P = E/t$, whereas t in s is the current drain time of discharging [1, 6].

2.3 Results and Discussion

2.3.1 *Structural Evolution*

The TRG was prepared by vacuum promoted thermal exfoliation of GO. Figure 2.1a shows a huge volume expansion obtained through the transformation from GO to TRG. In a typical procedure, ~0.6 g of TRG (G250) was produced by exfoliating 1.0 g of GO dry powder. Figure 2.1b, c show the SEM images of GO and G250. It can be seen that GO exhibits compact bulk morphology, while G250 owns a 3D honeycomb-like nanostructure. Numerous exterior macropores can be clearly identified among the lateral edges of graphene sheets. After high temperature annealing as shown in Fig. 2.1d, G1000 still maintained a honeycomb-like structure. However, a basal plane with a quite irregular edge is identified in G250 (Fig. 2.1e inset), while a more graphitic texture with a trimming edge can be found in G1000 (Fig. 2.1f inset), as indicated by the TEM.

N_2 adsorption isotherm, XRD and Raman spectroscopy were employed to investigate the morphological and structural evolution from graphite to TRG. As shown in Fig. 2.2a, all the isotherms give rise to a similar type III isotherm with H3 hysteresis loop in IUPAC classification, implying a unique adsorption behavior in the slit-shaped macropores raised from aggregates of plate-like particles. The specific BET surface area (S_{BET}), t-plot micropore surface area ($\Phi < 2$ nm) (S_{micro}), and total pore volume (V_P) of RGs are shown in Table 2.1. It is noteworthy that the S_{BET} is dramatically increased from 46.4 m^2 g^{-1} of precursor GO to 308.8 m^2 g^{-1} of G250, and the value keeps steady around 300 m^2 g^{-1} for G600 and G800, however raises to 434.1 m^2 g^{-1} for G1000. Besides, all the TRGs afford large pore volumes (V_P) of ~3 mL g^{-1}, which are higher than most of the ordered porous carbon materials (usually less than 1 mL g^{-1}) [31–34]. The BJH adsorption pore size distribution of G250 and G1000 (Fig. 2.2b) and t-plot micropore data (Table 2.1) indicate that the surface area and pore volume is mainly dependent on the pores with larger diameters ($\Phi > 2$ nm).

The XRD patterns of all the samples are given in Fig. 2.2c. The interlayer space (d_{002}) is increased from 0.334 nm of graphite to 0.698 nm of GO due to the efficient chemical intercalation by functional groups and water, which indicates an efficient transformation from GO to graphene. Furthermore, all TRG samples exhibit very weak diffraction peaks, owing to the dissociation of long-range ordered stacking of GO by random interlayer expansion and exfoliation [35].

The nature of disorder in GO and TRG is further studied by Raman spectroscopy (Fig. 2.2d). The in-plane vibration of a sp^2-hybridized carbon in the graphite crystallites (G band) as well as the disorder band aroused from amorphous carbon and

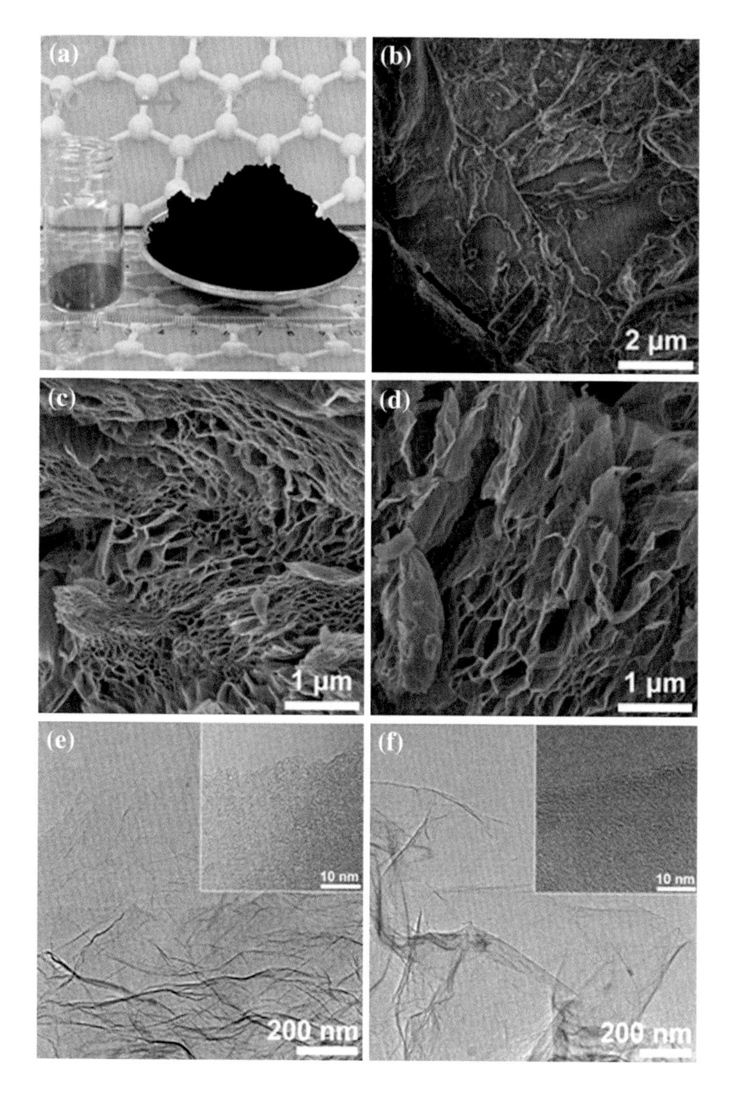

Fig. 2.1 **a** Digital images of the GO precursor (1.0 g) and resulted G250 (~600 mg); SEM images of **b** GO, **c** G250, and **d** G1000; TEM images of **e** G250 and **f** G1000. *Inserted figures* are the high resolution TEM images of **e** G250 and **f** G1000. Reprinted from Ref. [55], Copyright 2012, with permission from Elsevier

edges (D band) are identified, respectively. The peak positions and I_D/I_G ratios as calculated from Raman spectra (Fig. 2.2d) are provided in Table 2.1. Along the graphite-GO-TRG path, both the D and G band undergo significant changes upon amorphization of graphite, as amorphous carbon contains a certain fraction of sp^3 carbons. First, from graphite to GO, the G band widens significantly with a blue

Table 2.1 Summary of pore structure and elemental composition of TRG

Sample	D band[a] (cm^{-1})	G band[a] (cm^{-1})	I_D/I_G[a]	S_{BET}[b] ($m^2\ g^{-1}$)	S_{micro}[b] ($m^2\ g^{-1}$)	V_P[b] ($mL\ g^{-1}$)	C[c] (at.%)	O[c] (at.%)	C/O atom ratio[c]	Residual carbon[d] (%)
Graphite	1348.9	1574.5	0.16	–	–	–	–	–	–	–
GO	1336.7	1588.7	0.97	46.4	–	–	72.18	27.82	2.59	44.3
G250	1331.1	1591.5	1.42	308.8	8.34	3.578	89.70	10.30	8.71	62.2
G600	1330.7	1587.9	1.47	293.1	NA	2.905	91.19	8.81	10.35	71
G800	–	–	–	302.2	NA	3.192	95.34	4.66	20.46	79.7
G1000	1331.1	1594.8	1.44	434.1	NA	3.138	97.06	2.94	33.05	84.8

Reprinted from Ref. [55], Copyright 2012, with permission from Elsevier
[a]Obtained from Raman spectroscopy
[b]Calculated from N_2 physisorption isotherm
[c]Quantified by XPS
[d]Obtained from TGA (from room temperature to 1000 °C with rate of 10 °C/min in Ar)

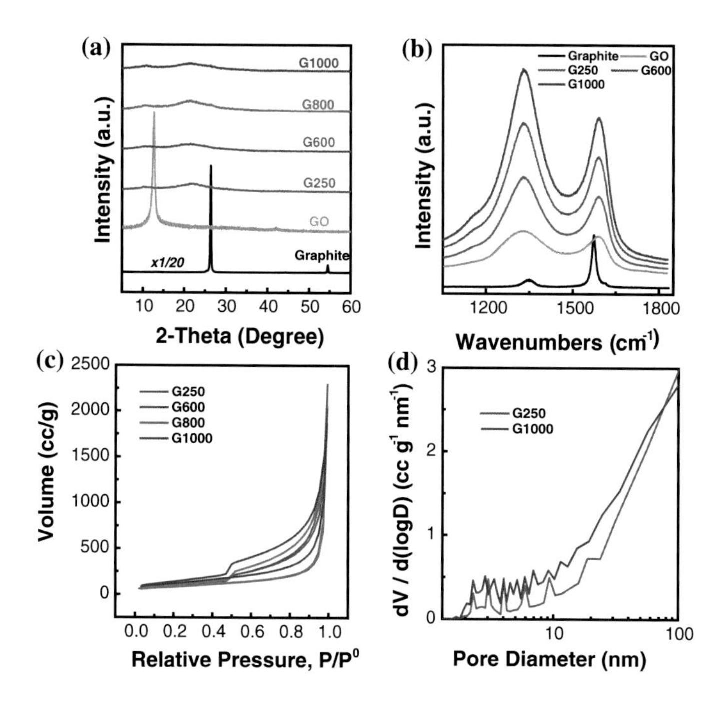

Fig. 2.2 **a** XRD pattern and **b** Raman spectrum evolution from GO to TRG; **c** N_2 physical adsorption isotherm and **d** BJH adsorption pore size distribution of G250 and G1000. Reprinted from Ref. [55], Copyright 2012, with permission from Elsevier

shift to higher frequencies of peak position and the intensity ratio of D and G bands increases, which can be attributed to some in-plane sp^2 carbons in graphitic domain transform into distorted sp^3 carbons in amorphous domain. Thus, a pristine graphitic region may be divided into several isolated sp^2 domains and double bonds by these amorphous domains, which resonate at higher frequencies than the G band of graphite. Besides, the newly introduced amorphous domains will also cause the increase in the intensity ratio of D and G band [36–39]. Second, a gradual sharpening of G band as well as increasing of I_D/I_G ratio (1.4–1.5) are observed from GO to G1000. During annealing, the desorption of oxygen bonded saturated sp^3 carbons as CO_2 and CO would leave various topological defects and vacancies in the graphene lattice, while a simultaneous 'self-healing' of graphitic lattice ('re-graphitization') will occur owing to the continuous deoxygenation.

FTIR spectrum is employed to characterize the surface functionalities of graphene qualitatively. As shown in Fig. 2.3, GO shows a wide hydroxyl stretching vibration mode in carboxyl, phenol, and/or intercalated H_2O (ν O–H) at 3400 cm^{-1}, and the C=O stretching vibration from carbonyl and carboxyl groups (ν C=O) at 1710 cm^{-1}. The characteristic peak at 1620 cm^{-1} corresponds to the components from the skeletal vibrations of un-oxidized graphitic domains

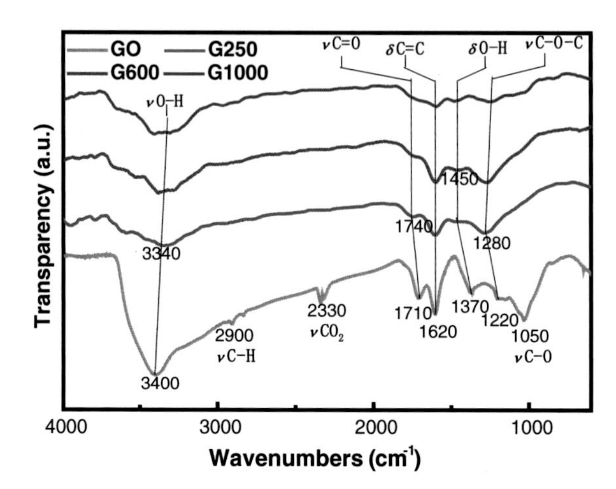

Fig. 2.3 The FTIR spectra of GO and TRG (G250, G600, and G1000). Reprinted from Ref. [55], Copyright 2012, with permission from Elsevier

(δ C=C). While the bands at 1370, 1220, and 1050 cm^{-1} can be linked to the carboxyl C–O deformation vibrations (δ O–H), epoxy, and/or ether type C–O–C (ν C–O–C), and alkoxy C–O stretching vibrations (δ C–O), respectively [40–42]. Besides, the thermal reduction temperature has a remarkable impact on the oxygen-containing groups. As the thermal reduction process deepened at higher temperature, the oxygen-containing groups are diminishing significantly. For G250 and G600, the ν O–H peak is blue shifted to 3340 cm^{-1} with a weaker intensity, while the ν C–O–C is red shifted to 1280 cm^{-1} with a relative higher intensity. Such phenomenon is ascribed to the dehydration of some neighboring –OH species, which not only weaken the intramolecular hydrogen bonding but also introduce more thermally stable C–O–C species. Moreover, the ν C=O peak was remarkably diminished due to the decomposing of thermally unstable –COOH species. The remained peak, up shifted to 1740 cm^{-1}, can be a trace of C=O groups in quinones, ketones, and lactones [43]. The vanishing of ν C–O peak implies the removal or transformation of oxygen saturated sp^3 carbon sites whereas the remained strong C=C mode is due to the effective repairing of conjugated C=C in sp^2 graphitic regions. Finally, G1000 shows a characteristic of infrared inert, which can be attributed to the increase of structural symmetry by intensive deoxygenation and thermal healing of graphitic domains at higher temperature.

The atomic percentage of C, O heteroatoms, as well as the C/O atomic ratio are calculated and summarized in Table 2.1. After thermal expansion of GO in high vacuum at 250 °C, the oxygen content significantly decreases to 10.30 at.% of G250 from 27.82 at.% of GO. As the annealing temperature increases, oxygen on the surface of TRGs are progressively removed with a final value of only 2.94 at.% for G1000 (Table 2.1), indicating a high degree of carbonization. The chemical state of

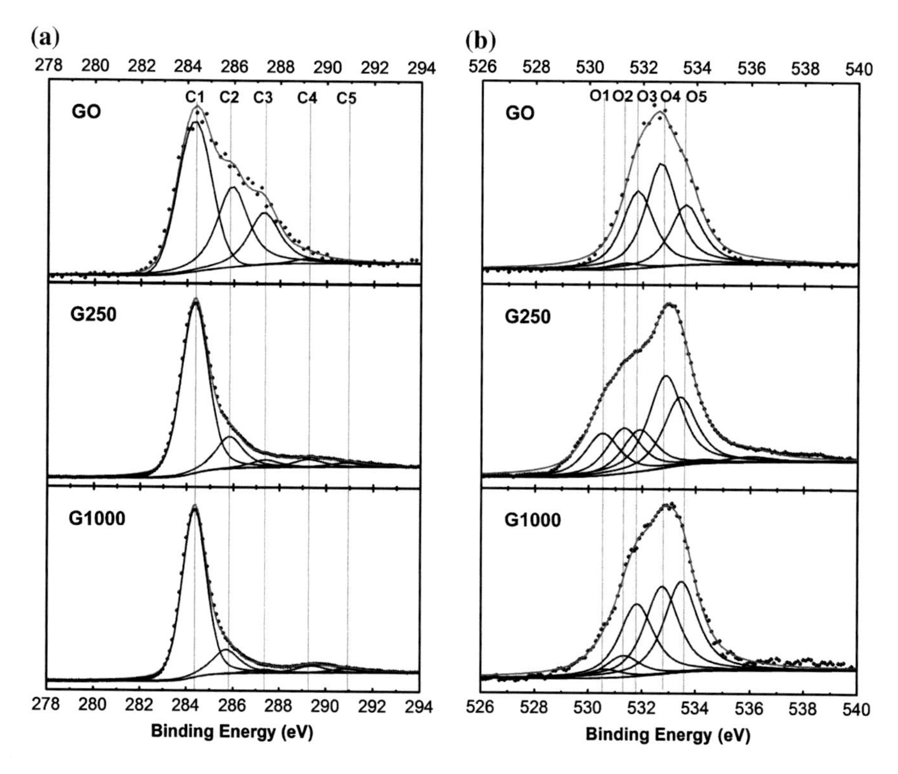

Fig. 2.4 **a** The C1s and **b** O1s XPS fine scan spectrum of GO, G250, and G1000. Reprinted from Ref. [55], Copyright 2012, with permission from Elsevier

the carbon and oxygen species on the surface of GO and TRGs was further studied by XPS analysis. As shown in Fig. 2.4a, b, various C (C1–C5) and O (O1–O5) components in GO, G250 and G1000 are determined by fitting the C1s and O1s fine scan spectra. C1 (284.4 eV), C2 (285.9 eV), C3 (287.4 eV), C4 (289.3 eV), and C5 (290.8 eV) components are belong to C–C, C–O, C=O, C(O)O related groups and graphitic shake-up satellites, respectively. O1 (530.4 eV), O2 (531.2 eV), O3 (531.9 eV), O4 (532.8 eV), and O5 (533.4 eV) components are assigned to quinone-type, C(O)O, –C=O, –C–O, and –OH groups, respectively [39, 44, 45].

As shown in Table 2.2, the amounts of C–C components are significantly increased from 33.75 at.% of GO to 73.13 at.% of G1000, with a simultaneous improvement of the graphitization degree, while the abundance of C–O and C=O related species are correspondingly decreased. Correlated with the Raman spectra and FTIR curves, the fitting results of C1s spectrum indicate that the oxygen heteroatoms are gradually removed from the basal plane of graphene during the thermal exfoliation and further annealing process. This will be favorable for the restoration of sp^2 conjugated carbon lattice so as to enhance the electronic conductivity of final products. Further, the O1s peak fitting results are more reliable in

Table 2.2 Fitted results (at.%) of C1s XPS spectra of TRG

B.E. (eV)	C1 (284.4)	C2 (285.9)	C3 (287.4)	C4 (289.3)	C5 (290.8)
Assignment	C–C	C–O	C=O	C(O)O	Graphitic shake-up
GO	33.75	22.57	14.70	1.16	0.00
G250	64.98	14.82	3.81	4.04	2.26
G1000	73.13	14.19	2.07	4.49	2.60

quantitating the relative accurate composition of different oxygen components in GO and TRGs. As shown in Table 2.3, the C=O components (mainly contributed by carboxyls and carbonyls) are remarkably reduced from 8.71 at.% of GO to 1.40 at.% of G250, while the C(O)O related components (mainly ascribed to anhydrides and lactones) are simultaneously increased from 0.66 of GO to 1.52 at.% of G250. Therefore, it is deduced that, a vast majority of unstable carboxylic groups in GO is either removed by thermal decarboxylation, or rather transformed into more thermally stable anhydrides and lactones during the vacuum promoted thermal expansion at 250 °C. It is worth mentioning that, about 1.41 at.% of quinone-type oxygens are also introduced to graphene after the thermal-induced exfoliation. Since some unadulterated free radicals will generate on the edge and basal plane of graphene during the elimination of functionalities, the oxygen atoms in the circumstance are tend to adsorb on these radicals to form the carbene-type coordinate intermediates. In order to give thermodynamically stable quinones, the rearrangement reactions take place between the delocalized π electrons from the basal plane of graphene and these coordinate intermediates [46, 47]. Meanwhile, a majority of C–O and O–H components are also eliminated by the intramolecular dehydration between hydroxylic and/or carboxylic groups. After carbonized at 1000 °C, most of the thermally unstable oxygen components in graphene have been thoroughly removed. As a result, the O–H, C–O, and C=O related functional groups, which are, respectively, transformed into thermally more stable isolated phenols, ethers, and carbonyls, finally become the major components in G1000.

The thermogravimetric analysis-mass spectrometry (TG-MS) is conducted in Ar atmosphere to examine the thermochemistry of GO and TRGs (Fig. 2.5). As shown in Table 2.1, after annealing at various temperatures from 250 to 1000 °C, the residual carbon ratios are increased progressively from 44.3 % of G250 to 84.8 % of G1000, which indicates the stabilization effect on graphene by thermal annealing. The sweep gases from TG are further analyzed by mass spectrometry and various fragments with m/z = 12.3 (radical C), 16.26 (radical O), 18.16 (H_2O), 28.18 (CO), and 44.09 (CO_2) are assigned (Fig. 2.5b–f). From GO to G1000, all the evolved gases fragments are dramatically decreased with the blue

Table 2.3 Fitted results (at.%) of O1s XPS spectra of TRG

B.E. (eV)	O1 (530.5)	O2 (531.2)	O3 (531.9)	O4 (532.7)	O5 (533.5)
Assignment	Quinone	C(O)O	C=O	C–O	O–H
GO	0.00	0.66	8.71	11.66	6.79
G250	1.41	1.52	1.40	3.09	2.28
G1000	0.11	0.27	0.90	1.09	1.13

shifts of decomposing temperature, which implies the improved stability of graphene at higher carbonization process.

The accompanying structural evolution and the surface chemistry of carbons have been widely investigated in the past several decades [47–49]. However, Graphene, as a single layer of sp^2 carbon atoms and two-dimensional carbon macromolecule, is giving new insights toward the surface chemistry of carbon materials [18, 46, 50, 51]. Since different oxygen-containing functional groups display distinct decomposing behavior during carbonization process, the surface properties of carbon are mainly dependent on the maximum temperature employed in the annealing process, as [52, 53]. It is reported that the sequence of thermal stability for various oxygen-containing groups on carbon surface and corresponding gases evolved during the decomposing are: carbonyl/quinones (CO) > lactones (CO_2) > phenols (CO) > anhydrides (CO and CO_2) > carboxylic acids (CO_2) [18, 52]. Herein, several evolution pathways for surface oxygen functionalities are proposed according to the FTIR, XPS, and TG-MS results in Fig. 2.6. The deoxygenation and reduction during annealing process is roughly divided into five stages: (I) from 30 to 150 °C, the physically adsorbed and intercalated water is removed ('physical' water); (II) from 150 to 400 °C, the intermolecular dehydration occur between neighboring carboxylic and/or hydroxyl groups, to transform into thermally stable lactones, anhydrides, ethers, and carbonyls as well as release abundant H_2O ('chemical' water). At the same time, the decarboxylation of individual carboxylic groups release abundant CO_2 [49]; (III) from 400 to 600 °C, the anhydrides decompose to release both CO_2 and CO, while the individual phenols decompose to evolve CO; (IV) from 600 to 800 °C, the lactones and individual ether desorb as CO_2 and CO, respectively, and (V) from 800 to 1000 °C, the most stable carbonyls and quinones will decompose to release CO [52].

2.3.2 *Electrochemical Performance*

The electrochemical performances of TRG-based supercapacitor electrodes were characterized by a three-electrode system in 6.0 M KOH aqueous electrolyte. Figure 2.7a shows typical CV curves of all TRGs at a scan rate of 3.0 mV s^{-1}. It can be observed that all of them exhibit the prominent capacitive behaviors, as the

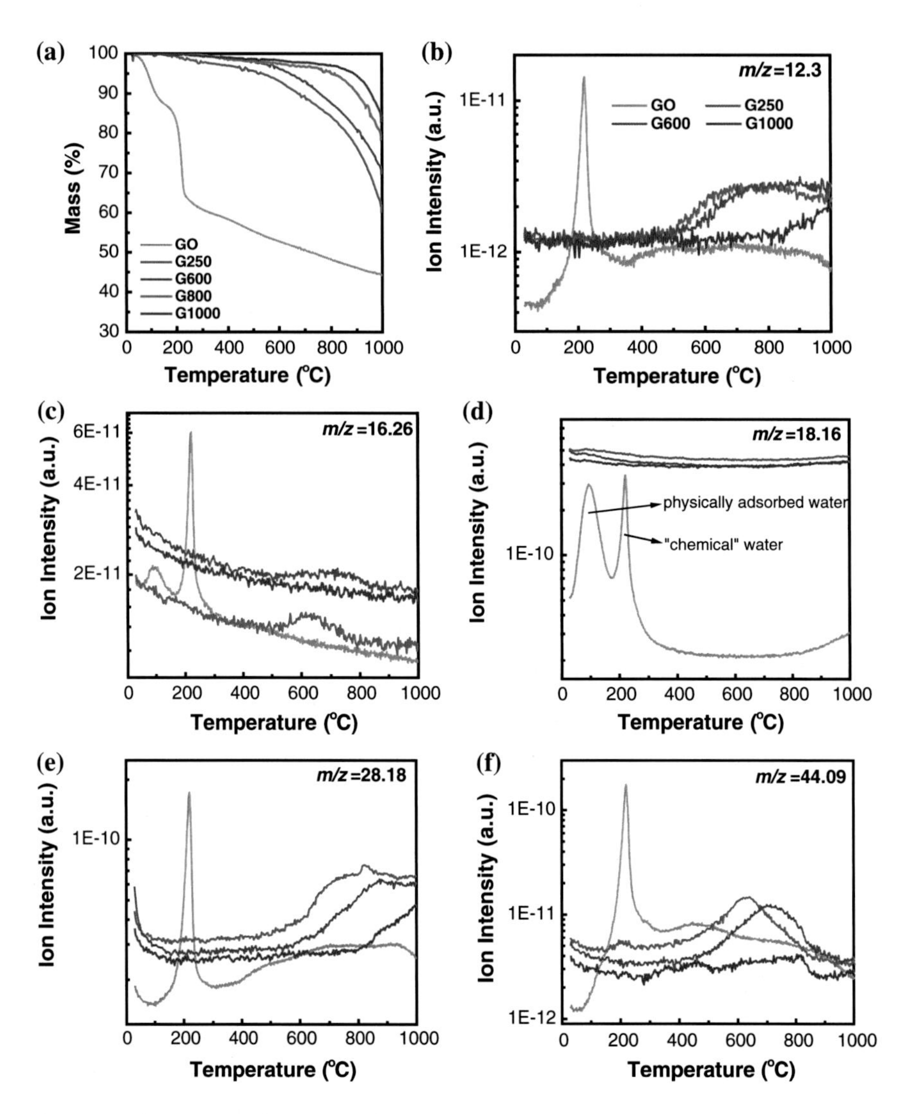

Fig. 2.5 **a–f** The TG-MS analysis of GO and TRGs. Reprinted from Ref. [55], Copyright 2012, with permission from Elsevier

sign of current reverse immediately upon reversal of the potential sweep. The capacitive current (i_M, A g^{-1}) drops quickly from G250 to G1000, while the CVs approach nearly ideal rectangular shape for, indicating an pronounced contribution of EDLC for the G1000-based electrode. The specific capacitance C_F (F g^{-1}) and C_s (F m^{-2}) as calculated from CVs (Fig. 2.7a) and S_{BET} (Table 2.1) are summarized in Fig. 2.7b. It can be seen that the initial CF and CS of G250 reach 170.5 F g^{-1} and 0.56 F m^{-2} at a low scan rate of 3 mV s^{-1}, respectively. However,

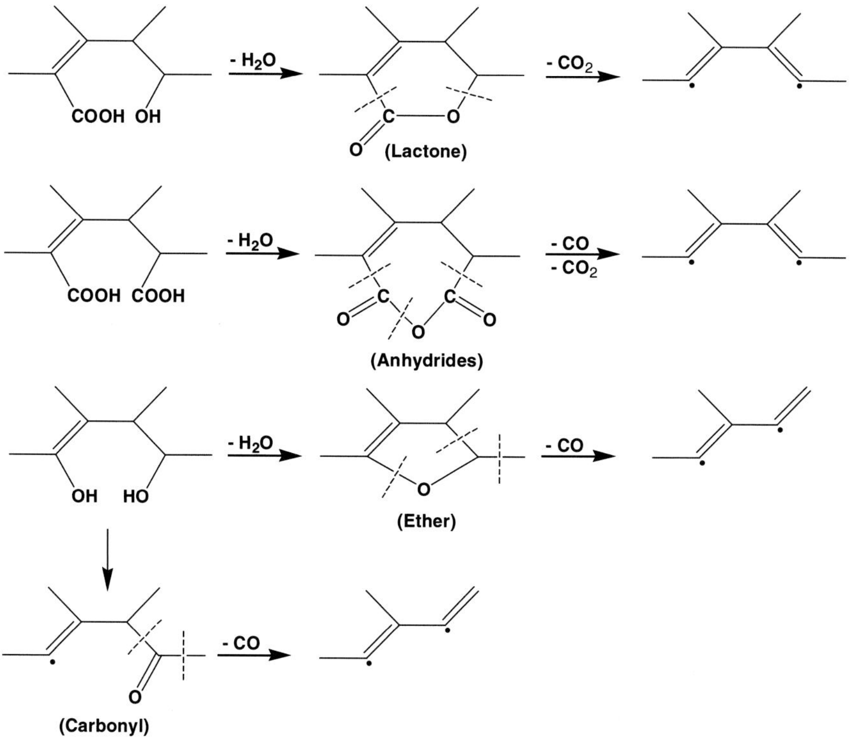

Fig. 2.6 Proposed pathway for transformation and evolution of oxygen-containing functional groups during thermal expansion of GO into TRGs. Reprinted from Ref. [55], Copyright 2012, with permission from Elsevier

the values decrease quickly for the further annealed TRGs. The G1000 presents a very poor specific capacitance ($C_F = 47.5$ F g^{-1}, $C_S = 0.11$ F m^{-2}). On one hand, most of the PC is eliminated owing to the intensive removal of PC-active oxygen-containing functional groups at higher temperature; on the other hand, the electronic conductivity of the electrode is significantly enhanced with the progressive re-graphitization of the carbon lattice in graphene basal plane.

In order to evaluate the fast charge/discharge abilities of the materials, the rate-dependent CVs for G250 and G1000 over a range of scan rates from 3 to 500 mV s^{-1} are examined as shown in Fig. 2.7c and d, respectively. When the scan rate increases, the CVs of G250 distort quickly to a willow-leaf-like shape with a voltage delay (ΔU) of ~0.2 V at 500 mV s^{-1}. This is attributed to the relatively slow charge/discharge kinetics from Faradaic reaction of surface functionalities compared with EDLC. At the same time, the CVs of G1000 still retain a rectangular-like shape and undergo much less distortion (ΔU ~0.06 V) at 500 mV s^{-1}, due to a quick response of EDLC with a pure electrostatic character. The specific capacitance values calculated from CVs at different scan rates are

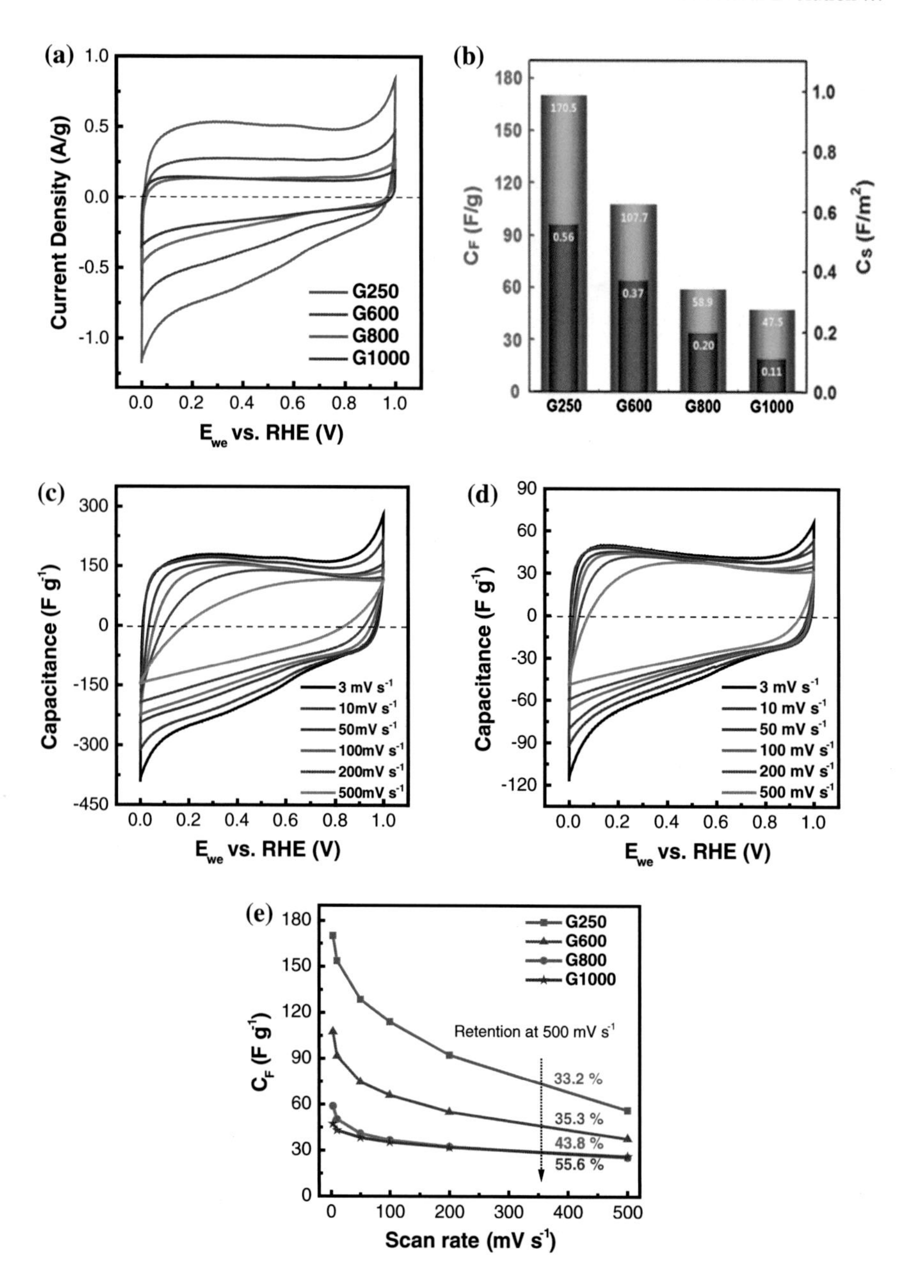

Fig. 2.7 **a** Evolution of CVs at 3.0 mV s^{-1} from G250 to G1000; CVs of **b** G250 and **c** G1000 with sweep rates from 3.0 to 500 mV s^{-1}; **d** Specific capacitance C_F (F g^{-1}) and C_s (F m^{-2}) for G250 to G1000 at 3.0 mV s^{-1}; **e** C_F versus sweep rates for G250 to G1000. Reprinted from Ref. [55], Copyright 2012, with permission from Elsevier

Table 2.4 Fitting results of internal components for graphene based electrodes

Sample	R_i (Ω)	R_c (Ω)	C_c (F/g)	W (Ω $s^{-1/2}$)	C_d (F/g)
G250	1.93	0.43	0.39	1.07	135.14
G1000	0.89	0.34	0.23	1.12	34.47

shown in Fig. 2.7e. Compared with the other TRGs, G250 exhibits the highest capacitive value within the whole range of scan rates. Although the initial C_F at 3 mV s^{-1} is remarkably reduced for the higher temperature annealed TRGs, the retention of C_F at a high scan rate of 500 mV s^{-1} is gradually increased from 33.2 % of G250 to 55.6 % of G1000. With the increase of scan rates, the EDLC becomes the primary electrochemical behavior of the electrodes owing to the delay of potential during reversing scan, which is related to a kinetically slow process involved during charging/discharging the PC [1].

Figure 2.8a clearly shows the GC curves of TRGs at the current density of 1.0 A g^{-1}. The C_F calculated from the discharge curves of G250, G600, G800, and G1000 are 167.3, 98.3, 49.4, and 42.6 F g^{-1}, while the voltage drop (IR_{drop}) at the initiation of the discharge is 9.6, 9.9, 8.1, and 8.7 mV, respectively, indicating a very low equivalent series resistance (ESR) in the test cell. Figure 2.8b exhibits Nyquist plots of G250 and G1000. Meanwhile, an equivalent circuit model in inset of Fig. 2.8b is introduced to simulate the capacitive and resistive elements of the cells under analysis, which contain the internal resistance of the TRG-based electrode (R_i), the capacitance and resistance due to contact interface (C_c and R_c), a Warburg diffusion element attributable to the ion migration through the graphene (Z_w), and the capacitance inside the pores (C_d) [54]. The corresponding results are presented in Table 2.4. In accordance with the microstructural and chemical evolution of graphene, the internal resistance (R_i) of G1000 (0.89 Ω) is remarkably lower than that of G250 (1.93 Ω), which is ascribed to the sp^2 conjugated carbon lattice significantly restored during the annealing process to increase charge carrier density. Moreover, comparing with G250 (Z_w = 1.07 Ω $s^{-1/2}$), G1000 exhibits a slightly higher Warburg diffusion resistance (Z_w = 1.12 Ω $s^{-1/2}$), which is attributed to the hamper of ion transfer between the inner channel of honeycomb and the aqueous electrolyte, as the wettability of the graphene surface is remarkably tampered after removal of most hydrophilic oxygen groups. Finally, the G250 electrode delivers the inner-porous capacitance of 135.1 F g^{-1}, whereas only about a quarter of the value of G250 (34.5 F g^{-1}) is obtained for the G1000 electrode, which imply oxygen functional groups play a crucial role in both introducing abundant peseudocapacitance as well as improving the surface efficiency of graphene during electrochemical cycling.

The contribution of pseudocapacitance from oxygen functionalities is based on a reversible Faradic redox mechanism [1, 3]. The basal plane of a TRG sheet is composed of distorted sp^3 carbons decorated by various oxygen functionalities and numerous sp^2 graphitic domains [50]. The inductive effects originated from the

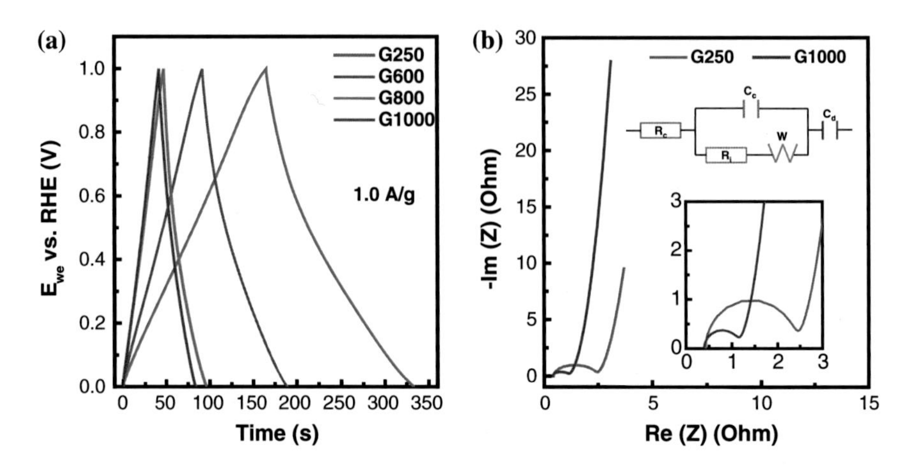

Fig. 2.8 **a** Galvanostatic charge/discharge curves and **b** Nyquist plot of G250 and G1000; *Inserted figure* show the enlarged Nyquist plot and equivalent circuit model. Reprinted from Ref. [55], Copyright 2012, with permission from Elsevier

oxygen atoms will lead to the polarization of some C–C bonds, while the delocalized electrons within the sp^2 domains supply a low resistant channel for charge transfer within the electrode. Thus, the reversible redox reactions are likely to occur on these polarized sites during charging/discharging process (Fig. 2.9).

Comparing with basic functional groups (Fig. 2.9a, d), it is believed that the acidic sites (Fig. 2.9b, c) on carbon play a more dominant role in the alkaline KOH electrolyte. The abundant carboxylic and phenolic groups, which are usually thermally unstable, can be well maintained at a mild annealing temperature. However, with the temperature increasing, these oxygen-containing functional groups will be removed or transformed into basic sites, which are adverse to the pseudocapacitive reactions. This is the reason that the mild annealed G250, which not only possesses the highest oxygen content, but also own abundant acidic sites, shows extremely high electrochemical performance as supercapacitor electrode materials.

In order to completely determine the electrochemical performance of TRGs, a Ragone plot is exhibited in Fig. 2.10a. The energy and power densities are calculated from GC curves of a supercapacitor with a cell-voltage window of 1.0 V. It is worth noting that G250 exhibits the best electrochemical performance in energy storage among the TRGs-based electrodes, with a maximum energy density of 11.6 Wh kg^{-1} corresponding to power density of 250 W kg^{-1}. These results are in good agreement with the capacitance performance as calculated by CV curves. When the power density reaches as high as 5000 W kg^{-1}, it still owns an energy density of no less than 7.8 Wh kg^{-1}.

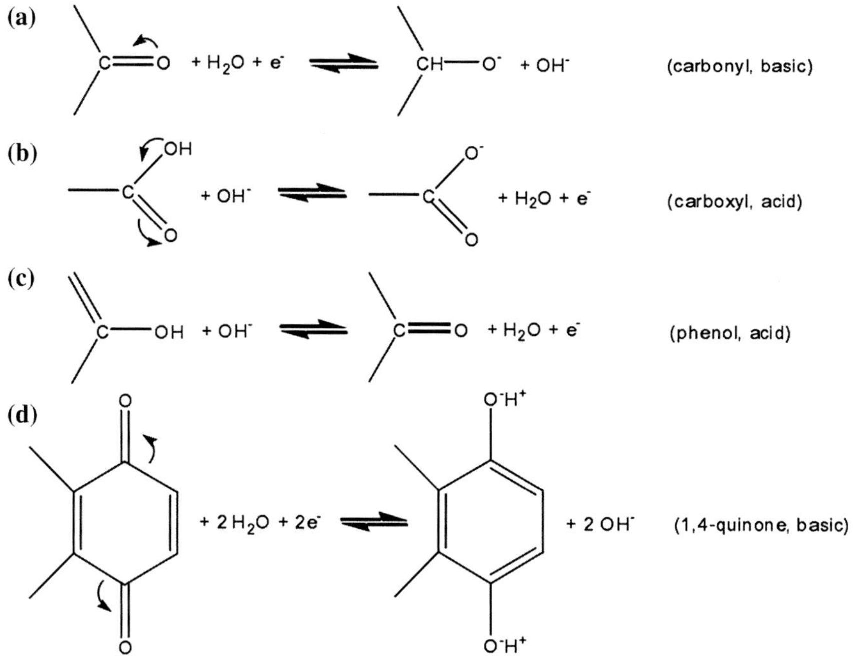

Fig. 2.9 Proposed reversible pseudo-capacitive reactions of various oxygen-containing functional sites with its acid-base property in the aqueous electrolyte. Reprinted from Ref. [55], Copyright 2012, with permission from Elsevier

In order to evaluate the cycling stability of G250 and G1000 electrodes, repeating CV charge/discharge process were performed at a scan rate of 50 mV s^{-1} between 0 V and 1 V in 6 M KOH aqueous electrolyte. As shown in Fig. 2.10b, after 3000 cycles, the specific capacitance of G250 was reduced to 94.7 % of the initial cycle, while G1000 still retains a high capacitive retention of as high as 99.5 %. The disparity of cycling stability between G250 and G1000 is ascribed to the difference in surface chemistry for TRGs at different carbonation stages. Comparing with G1000, G250 owns a much higher density of oxygen functional groups, especially for the unstable acidic carboxyls, which may be neutralized and diminished by alkaline KOH during repeating charging/discharging cycles, resulting in the loss of PC. Overall, comparing with the other energy storage devices which often suffer from a short cycle life due to irreversible physical and/or chemical changes; TRGs-based supercapacitor is still very attractive for its prominent durability and stability.

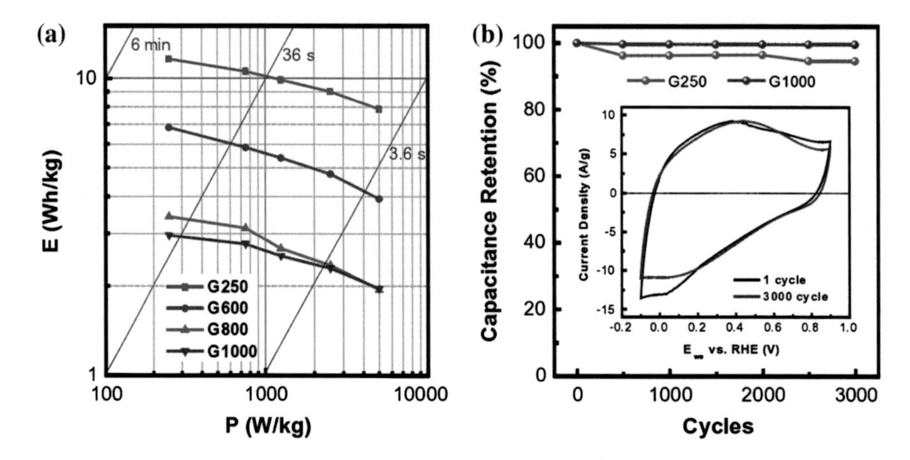

Fig. 2.10 Energy storage performance of TRGs: **a** Ragone plot as calculated from GC results under current density of 1, 3, 5, 10 and 20 A g^{-1}; **b** Cycling performance of G250 and G1000. *Inset* shows the 1 and 3000 cycle CV of G250 at a scan rate of 50 mV s^{-1} in 6.0 M KOH. Reprinted from Ref. [55], Copyright 2012, with permission from Elsevier

2.4 Conclusions

In this study, thermally reduced graphene (TRG) with adjustable surface chemistry was prepared by vacuum promoted thermal expansion of GO and subsequent annealing at different temperatures. After thermal expansion of GO in high vacuum at 250 °C, the C=O components (mainly ascribed to carboxyls and carbonyls) are dramatically decreased from 8.71 at.% of GO to only 1.40 at.% of G250, whereas the C(O)O related components (mainly contributed by anhydrides and lactones) are simultaneously increased from 0.66 at.% of GO to 1.52 at.% of G250. After carbonization at 1000 °C, most oxygen components in TRG have been removed. The O–H, C–O, and C=O related functional groups, which belong respectively to thermally more stable isolated phenols, ethers, and carbonyls, finally become the major components in G1000. Besides, the desorption of oxygen bonded carbons as CO_2 and CO produces large amount of topological defects and vacancies in the graphene lattice, with a simultaneous 'self-healing' of graphitic lattice ('re-graphitization'). The TRGs were further applied as electrodes of supercapacitor. The studies showed that the structural evolution induced by thermal annealing plays a key role in determining the electrochemical performance of TRGs towards supercapacitor applications. The oxygen functional groups can significantly enhance the capacitance performance of TRGs by introducing abundant PC-active sites through reversible Faradic redox reactions. It is worth noting that no obvious capacitance loss is observed over 3000 charge/discharge cycles, clearly demonstrating good cycling durability. TRGs-based supercapacitor will benefit the application of graphene in advanced energy storage.

References

1. Dell RM, Rand DAJ. Energy storage-a key technology for global energy sustainability. J Power Sources. 2001;100:2–17.
2. Conway BE. Electrochemical supercapacitors: scientific fundamentals and technological apllication. Dordrecht: Kluwer Academic/Plenum; 1999.
3. Frackowiak E. Carbon materials for supercapacitor application. Phys Chem Chem Phys. 2007;9(15):1774–85.
4. Pandolfo AG, Hollenkamp AF. Carbon properties and their role in supercapacitors. J Power Sources. 2006;157(1):11–27.
5. Su DS, Schlogl R. Nanostructured carbon and carbon nanocomposites for electrochemical energy storage applications. ChemSusChem. 2010;3(2):136–68.
6. Kotz R, Carlen M. Principles and applications of electrochemical capacitors. Electrochim Acta. 2000;45(15–16):2483–98.
7. An SJ, Zhu YW, Lee SH, Stoller MD, Emilsson T, Park S, et al. Thin film fabrication and simultaneous anodic reduction of deposited graphene oxide platelets by electrophoretic deposition. J Phys Chem Lett. 2010;1(8):1259–63.
8. Zhu Y, Murali S, Stoller MD, Ganesh KJ, Cai W, Ferreira PJ, et al. Carbon-based supercapacitors produced by activation of graphene. Science. 2011;332(6037):1537–41.
9. Su FY, You CH, He YB, Lv W, Cui W, Jin FM, et al. Flexible and planar graphene conductive additives for lithium-ion batteries. J Mater Chem. 2010;20(43):9644–50.
10. Wu Z-S, Ren W, Wen L, Gao L, Zhao J, Chen Z, et al. Graphene anchored with Co_3O_4 nanoparticles as anode of lithium ion batteries with enhanced reversible capacity and cyclic performance. ACS Nano. 2010;4(6):3187–94.
11. Wu Z-S, Wang D-W, Ren W, Zhao J, Zhou G, Li F, et al. Anchoring hydrous RuO_2 on graphene sheets for high-performance electrochemical capacitors. Adv Funct Mater. 2010;20(20):3595–602.
12. McAllister MJ, Li J-L, Adamson DH, Schniepp HC, Abdala AA, Liu J, et al. Single sheet functionalized graphene by oxidation and thermal expansion of graphite. Chem Mater. 2007;19(18):4396–404.
13. Zhu Y, Stoller MD, Cai W, Velamakanni A, Piner RD, Chen D, et al. Exfoliation of graphite oxide in propylene carbonate and thermal reduction of the resulting graphene oxide platelets. ACS Nano. 2010;4(2):1227–33.
14. Lv W, Tang D-M, He Y-B, You C-H, Shi Z-Q, Chen X-C, et al. Low-temperature exfoliated graphenes: vacuum-promoted exfoliation and electrochemical energy storage. ACS Nano. 2009;3(11):3730–6.
15. Zhang HB, Wang JW, Yan Q, Zheng WG, Chen C, Yu ZZ. Vacuum-assisted synthesis of graphene from thermal exfoliation and reduction of graphite oxide. J Mater Chem. 2011;21(14):5392–7.
16. Wu Z-S, Ren W, Gao L, Zhao J, Chen Z, Liu B, et al. Synthesis of graphene sheets with high electrical conductivity and good thermal stability by hydrogen arc discharge exfoliation. ACS Nano. 2009;3(2):411–7.
17. Wu Z-S, Ren W, Gao L, Liu B, Jiang C, Cheng H-M. Synthesis of high-quality graphene with a pre-determined number of layers. Carbon. 2009;47(2):493–9.
18. Bagri A, Mattevi C, Acik M, Chabal YJ, Chhowalla M, Shenoy VB. Structural evolution during the reduction of chemically derived graphene oxide. Nat Chem. 2010;2(7):581–7.
19. Pumera M. Graphene-based nanomaterials and their electrochemistry. Chem Soc Rev. 2010;39(11):4146–57.
20. Yan J, Wei T, Shao B, Fan ZJ, Qian WZ, Zhang ML, et al. Preparation of a graphene nanosheet/polyaniline composite with high specific capacitance. Carbon. 2010;48(2):487–93.
21. Wang HL, Hao QL, Yang XJ, Lu LD, Wang X. A nanostructured graphene/polyaniline hybrid material for supercapacitors. Nanoscale. 2010;2(10):2164–70.

22. Zhang LL, Zhou R, Zhao XS. Graphene-based materials as supercapacitor electrodes. J Mater Chem. 2010;20(29):5983–92.
23. Sun YQ, Wu QO, Shi GQ. Graphene based new energy materials. Ener Environ Sci. 2011;4(4): 1113–32.
24. Fan ZJ, Yan J, Zhi LJ, Zhang Q, Wei T, Feng J, et al. A three-dimensional carbon nanotube/ graphene sandwich and its application as electrode in supercapacitors. Adv Mater. 2010;22(33): 3723–8.
25. Huang X, Yin ZY, Wu SX, Qi XY, He QY, Zhang QC, et al. Graphene-based materials: synthesis, characterization, properties, and applications. Small. 2011;7(14):1876–902.
26. Guo SJ, Dong SJ. Graphene nanosheet: synthesis, molecular engineering, thin film, hybrids, and energy and analytical applications. Chem Soc Rev. 2011;40(5):2644–72.
27. Inagaki M, Konno H, Tanaike O. Carbon materials for electrochemical capacitors. J Power Sources. 2010;195(24):7880–903.
28. Zhu YW, Murali S, Cai WW, Li XS, Suk JW, Potts JR, et al. Graphene and graphene oxide: synthesis, properties, and applications. Adv Mater. 2010;22(35):3906–24.
29. Fan ZJ, Yan J, Wei T, Zhi LJ, Ning GQ, Li TY, et al. Asymmetric supercapacitors based on graphene/MnO_2 and activated carbon nanofiber electrodes with high power and energy density. Adv Funct Mater. 2011;21(12):2366–75.
30. Chen CM, Yang Q-H, Yang Y, Lv W, Wen Y, Hou P-X, et al. Self-assembled free-standing graphite oxide membrane. Adv Mater. 2009;21(29):3007–11.
31. Vix-Guterl C, Frackowiak E, Jurewicz K, Friebe M, Parmentier J, Béguin F. Electrochemical energy storage in ordered porous carbon materials. Carbon. 2005;43(6):1293–302.
32. Vinu A, Ariga K, Mori T, Nakanishi T, Hishita S, Golberg D, et al. Preparation and characterization of well-ordered hexagonal mesoporous carbon nitride. Adv Mater. 2005;17(13):1648–52.
33. Zhao XC, Wang AQ, Yan JW, Sun GQ, Sun LX, Zhang T. Synthesis and electrochemical performance of heteroatom-incorporated ordered mesoporous carbons. Chem Mater. 2010;22(19):5463–73.
34. Huang CH, Doong RA, Gu D, Zhao DY. Dual-template synthesis of magnetically-separable hierarchically-ordered porous carbons by catalytic graphitization. Carbon. 2011;49(9):3055–64.
35. Schniepp HC, Li J-L, McAllister MJ, Sai H, Herrera-Alonso M, Adamson DH, et al. Functionalized single graphene sheets derived from splitting graphite oxide. J Phys Chem B. 2006;110(17):8535–9.
36. Stankovich S, Dikin DA, Piner RD, Kohlhaas KA, Kleinhammes A, Jia Y, et al. Synthesis of graphene-based nanosheets via chemical reduction of exfoliated graphite oxide. Carbon. 2007;45(7):1558–65.
37. Pei SF, Zhao JP, Du JH, Ren WC, Cheng HM. Direct reduction of graphene oxide films into highly conductive and flexible graphene films by hydrohalic acids. Carbon. 2010;48(15):4466–74.
38. Kudin KN, Ozbas B, Schniepp HC, Prud'homme RK, Aksay IA, Car R. Raman spectra of graphite oxide and functionalized graphene sheets. Nano Lett. 2007;8(1):36–41.
39. Chen CM, Huang JQ, Zhang Q, Gong WZ, Yang QH, Wang MZ, et al. Annealing a graphene oxide film to produce a free standing high conductive graphene film. Carbon. 2012;50(2):659–67.
40. Stankovich S, Piner RD, Nguyen ST, Ruoff RS. Synthesis and exfoliation of isocyanate-treated graphene oxide nanoplatelets. Carbon. 2006;44(15):3342–7.
41. Xiao J, Mei D, Li X, Xu W, Wang D, Graff GL, et al. Hierarchically porous graphene as a lithium–air battery electrode. Nano Lett. 2011;11(11):5071–8.
42. Lai LF, Chen LW, Zhan D, Sun L, Liu JP, Lim SH, et al. One-step synthesis of NH_2-graphene from in situ graphene-oxide reduction and its improved electrochemical properties. Carbon. 2011;49(10):3250–7.
43. Yue L, Li W, Sun F, Zhao L, Xing L. Highly hydroxylated carbon fibres as electrode materials of all-vanadium redox flow battery. Carbon. 2010;48(11):3079–90.

44. Arrigo R, Havecker M, Wrabetz S, Blume R, Lerch M, McGregor J, et al. Tuning the acid/base properties of nanocarbons by functionalization via amination. J Am Chem Soc. 2010;132(28): 9616–30.
45. Yan J, Wei T, Shao B, Ma FQ, Fan ZJ, Zhang ML, et al. Electrochemical properties of graphene nanosheet/carbon black composites as electrodes for supercapacitors. Carbon. 2010;48(6): 1731–7.
46. Radovic LR. Active sites in graphene and the mechanism of CO_2 formation in carbon oxidation. J Am Chem Soc. 2009;131(47):17166–75.
47. Menendez JA, Xia B, Phillips J, Radovic LR. On the modification and characterization of chemical surface properties of activated carbon: microcalorimetric, electrochemical, and thermal desorption probes. Langmuir. 1997;13(13):3414–21.
48. Serp P, Figueiredo JL. Surface chemistry of carbon materials. In: Carbon materials for catalysis (Chap. 2), p. 45.
49. Menéndez JA, Phillips J, Xia B, Radovic LR. On the modification and characterization of chemical surface properties of activated carbon: in the search of carbons with stable Basic properties. Langmuir. 1996;12(18):4404–10.
50. Dreyer DR, Park S, Bielawski CW, Ruoff RS. The chemistry of graphene oxide. Chem Soc Rev. 2010;39(1):228–40.
51. Loh KP, Bao Q, Ang PK, Yang J. The chemistry of graphene. J Mater Chem. 2010;20(12): 2277.
52. Figueiredo JL, Pereira MFR. The role of surface chemistry in catalysis with carbons. Catal Today. 2010;150(1–2):2–7.
53. Acik M, Lee G, Mattevi C, Pirkle A, Wallace RM, Chhowalla M, et al. The role of oxygen during thermal reduction of graphene oxide studied by infrared absorption spectroscopy. J Phys Chem C. 2011;115(40):19761–81.
54. Xu GH, Zheng C, Zhang Q, Huang JQ, Zhao MQ, Nie JQ, et al. Binder-free activated carbon/ carbon nanotube paper electrodes for use in supercapacitors. Nano Res. 2011;4(9):870–81.
55. Chen CM, Zhang Q, Yang MG, Huang CH, Yang YG, Wang MZ. Structural evolution during annealing of thermally reduced graphene nanosheets for application in supercapacitors. Carbon. 2012;50(10):3572–84.

Chapter 3
Hierarchical Amination of Graphene for Electrochemical Energy Storage

3.1 Introduction

Recently, three-dimensional (3D) hierarchical architectures of nanosheets, nanoplates, nanotubes, nanowires, and nanospheres have attracted great interest in energy conversion and storage, nano-composites, sustainable catalysis, optoelectronics, and drug delivery systems, due to their outstanding electrochemical performance such as its ultrahigh surface-to-volume ratio, high porosity, strong mechanical strength, excellent electrical conductivity and fast mass, and electron transport kinetics [1, 2]. For example, various nanosheets, such as graphene and graphene oxide [3–9], layered double hydroxides [10], and natural clays [11], have been successfully applied in energy conversion and storage. Among various energy storage routes, supercapacitors are receiving increasing attention for complementing batteries in hybrid electric vehicles, portable electronics, and industrial power management owing to their large power density and longer life cycle [12–15]. The crucial issue for the successful application of hierarchical architectures in energy storage rests with the ability to manipulate the arrangement of the building blocks into well-designed architectures. For this purpose, it is still a great challenge to select building blocks with high surface area for electrical double-layer capacitance (EDLC), tailored surface chemistry for pseudocapacitance (PC), accessible macropores for ion-buffering reservoirs, and low-resistant pathways for the charge carrier transport. Graphene is a single layer of carbon atoms with a hexagonal arrangement in a two-dimensional lattice. Its high thermal conductivity, high specific surface area, excellent electronic conductivity, and huge theoretical surface area (2630 $m^2\ g^{-1}$) make it promising for potential applications in supercapacitor electrodes [16–21]. Graphene sheets derived from exfoliation of the graphite oxide formed by chemical oxidization of graphite are intrinsically decorated by abundant active sites, such as functional groups (mainly –OH, –COOH, –C=O, –C–O–C–) and lattice defects (atom vacancy, distortion, dangling bonds) on the lateral surface and at their edges [22–24]. The presence of functional groups on graphene is essential with respect to

C.-M. Chen, *Surface Chemistry and Macroscopic Assembly of Graphene for Application in Energy Storage*, Springer Theses,
DOI 10.1007/978-3-662-48676-4_3

improving the interfacial adhesion of rGO to active species. Many active species such as conductive polymers (e.g., PANI) [25, 26], metal oxide (e.g., RuO_2 [8], and MnO_2 [9]) have been introduced to improve the PC as well as the whole performance of graphene-based supercapacitor electrodes. However, such hybrids usually deliver poor rate capability owing to the large volumetric change or degradation during charge/discharge process. Therefore, it is still a big challenge to explore the basic building blocks of graphene with large EDLC and PC due to fast and reversible surface redox processes between the electrolyte and various electroactive species on graphene electrode surface. Generally, heteroatom modification has been proven to be the most promising method for enhancing the capacity, surface wettability of materials, and electronic conductivity, while maintaining a good cycling performance [27], for instance, superior supercapacitor performance was available on heteroatom-modified nanocarbon [27, 28]. Recently, the individual graphene sheets can also be designed into various kinds of macroscopic hierarchical architectures (such as graphene oxide–graphene film [29–34], hydrogel [35], and foam [36–38]) with high electronic conductivity, large surface area, proper pore distribution, and superior mechanical and chemical properties [29–38].

Based on this consideration, the hierarchically aminated graphene honeycombs (AGHs) have been explored by the exfoliation of bulk graphite oxide (GO) under high vacuum and subsequent amination of which at different temperatures. The microstructure and surface functionalities of AGHs were determined by high-angle annular dark-field (HAADF)-scanning transmission electron microscopy (STEM), electron energy loss spectroscopy (EELS), X-ray photoelectron spectroscopy (XPS), and thermogravimetry–mass spectroscopy (TG–MS). The results demonstrate that the as-formed AGHs exhibited large surface area, high charge carrier density, abundant PC-active sites with reversible Faradaic redox kinetics, and hierarchical 3D pores. When working as electrodes for supercapacitors, the AGHs exhibit high specific capacitance of 0.84 F m^{-2}, good rate performance, and outstanding cycle stability with capacitance retention of 97.8 % after 3000 cycles.

3.2 Experimental

3.2.1 Preparation of Hierarchically Aminated Graphene

The chemically derived graphene was obtained by low-temperature expansion of graphite oxide (GO) in high vacuum. GO was produced by a modified Hummers method [31]. The as-produced GO was loaded into a quartz tube, with one end sealed while the other end connected to a vacuum pump through a valve. The tube was pre-evacuated to the pressure of less than 2.0 Pa, and then a heating schedule with a heating rate of 30 °C min^{-1} was executed. An abrupt expansion of GO was observed with mass fluffy black powder generated at about 195 °C. To eliminate the superabundant functional groups, the GO was dwelled at 250 °C for 30 min in high vacuum (below 5 Pa) to obtain G250. The G250 was placed uniformly in a quartz boat, and put into the center of a quartz tube (Φ 80 mm, L 1200 mm). The quartz tube was kept at 200 °C with NH_3 flowing (200 mL min^{-1}) for 4.0 h to obtain AGH200. At the same time, AGH400 and AGH600 were also prepared by the similar procedures, with the only difference in amination temperature (400 and 600 °C, respectively).

Typically, 200 mL of the GO dispersion (1 mg mL^{-1}) was first sonicated (200 W) for 30 min to yield a homogeneous brown solution. Subsequently, the above solution was mixed with hydrazine hydrate (50 mL, 100 %) in a 500-mL round-bottom flask, and heated in an oil bath at 100 °C under a water-cooled condenser for 24.0 h over which the reduced GO gradually precipitated out as a black solid. After this, the product was separated by vacuum filtration and rinsed with distilled water (5 × 100 mL) and ethanol (5 × 100 mL) several times. Finally, the as-prepared sample was dried on the funnel under a continuous air flow through the solid product cake. The as-obtained control sample was named CRG.

3.2.2 Sample Characterization

The as-obtained G250, AGH200, AGH400, and AGH600 were characterized using scanning electron microscopy (SEM, Hitachi S4800, 2.0 kV), transmission electron microscope (TEM, Philips CM200, 200.0 kV), HAADF-STEM and EELS analyses (HAADF-STEM and EELS, Titan 80-300, 80.0 kV), and energy dispersive X-ray spectroscopy (EDX) analysis. Raman spectrophotometer (Renishaw RM2000 with laser excitation line at 633.0 nm), N_2 sorption isotherm, X-ray photoelectron spectroscopy (XPS, PHI Quantera SXM), and thermal gravimetric analysis (TGA) system (STA 409PC luxx, Netzsch, Germany) were equipped with a sweep gas mass spectroscopy appliance (ALZERS OmniStar 20, Switzerland).

3.2.3 Electrochemical Measurements

The working electrodes were prepared by mixing the active material, carbon black, and poly(tetrafluoroethylene) (PTFE) with a weight ratio of 80:5:15. A small amount of ethanol was added into the above mixture to form slurry and to form homogeneous slurry. The slurry was squeezed into a film and then punched into pellets (area ~0.8 cm^2). The punched pellets with a piece of nickel foam on each side were pressed under 2.5 MPa and dried overnight at 110 °C. Each electrode was quantified to contain ~3.5 mg active materials. A three-electrode experimental cell was equipped with a working electrode, a platinum foil counter electrode, and a saturated hydrogen reference electrode (SHE). Before testing, the electrodes were saturated with the electrolyte by vacuum enhanced impregnation for 2.0 h. Al electrochemical measurement was carried out in 6.0 M KOH aqueous electrolyte. Cyclic-voltammetry

(CV) curves and electrochemical impedance spectroscopy (EIS) profiles (frequency from 200 to 10 mHz) were measured with a VSP Bio-Logic electrochemistry workstation. The specific capacitance of the electrode (C_F, F g^{-1}) was calculated from CV curves by the following equation:

$$C_F = \frac{1}{sm\Delta V} \int_{V_0}^{V_o+\Delta V} i dV$$

where s in mV s^{-1} is the scan rate, m in g is the mass of active materials in the electrode, ΔV in V is the potential window (1 V in this work), V_0 in V is the initial value of potential window (0 V in this work), and i in A is the real time current during scanning. The Nyquist plots were fitted by EC-Lab software with the equivalent circuit as $R_c + C_c/(R_i + W) + C_d$, and the randomize plus Levenberg–Marquardt method was employed for the fitting. The Ragone plot was calculated from CV results, and the energy density (E, Wh kg^{-1}) was calculated by equation: $E = 1/2(C_F/4)\Delta V^2$, while power density (P, W kg^{-1}) was calculated by $P = E/t$, where t(s) is the current drain time of discharging [39, 40].

3.3 Results and Discussion

3.3.1 The Microstructure of AGHs

Comparing with the normal thermal expansion approach, the initial graphene honeycomb G250 can be obtained through the promotion of an external high vacuum environment [41]. Parameters on pore structure and elemental composition of G250, AGH200, AGH400, AGH600, and CRG are summarized in Table 3.1. Abundant oxygen functional groups are kept during exfoliation process, and the G250 retains a high oxygen content of 10.74 at.%. The G250 is further treated at 200, 400, and 600 °C for 4 h under flowing ammonia, and the as-obtained hierarchically aminated graphene honeycombs are named AGH200, AGH400, and AGH600, respectively. The SEM images of AGH200 are shown in Fig. 3.1a and b. The N-modified graphene sheets interact with each other to form a continuous and interconnected 3D macroscopic honeycomb, with a specific BET area of ~247 m^2 g^{-1}. Large porous network with size ranging from 30 to 200 nm is clearly visible among the lateral edges of multilayer graphene. Most of the graphene sheets are highly curved with many ripples and wrinkles on their surfaces. It is believed that the unique macroscopic honeycomb-like structure can be generated during thermal exfoliation process, where a negative pressure was surrounding the bulk graphite oxide to provide a moderate interlayer expansion. Furthermore, these pores can be well preserved during low-temperature (200 °C) amination for hierarchical AGHs.

The microtexture and chemical construction of the basic building blocks of AGH200 were determined by TEM, HAADF-STEM, and EDX elemental mapping techniques. The TEM and STEM images (Fig. 3.1c and d) show that the graphene

Table 3.1 Parameters on pore structure and elemental composition

Sample	$S_{BET}{}^{a}$ (m^2 g^{-1})	$S_{micro}{}^{a}$ (m^2 g^{-1})	$V_P{}^{a}$ (cm^3 g^{-1})	C^b (at.%)	O^b (at.%)	N^b (at.%)	$O + N^b$ (at.%)	Residual carbon[c] (wt%)
G250	293.2	3.663	3.58	89.26	10.74	0.00	10.74	62.6
AGH200	247.0	NA	3.40	86.91	10.30	2.79	13.09	71.1
AGH400	703.5	NA	3.32	87.87	8.39	3.74	12.13	84.3
AGH600	503.0	NA	3.17	90.86	5.23	3.91	9.14	91.1
CRG	666.0	69.93	0.55	88.43	11.57	0.00	10.74	–

Reproduced from Ref. [63] by permission of The Royal Society of Chemistry
[a]Obtained from N_2 adsorption–desorption isotherm
[b]Quantified by XPS
[c]Obtained from TGA (from room temperature to 1000 °C with a rate of 10 °C min^{-1} in Ar)

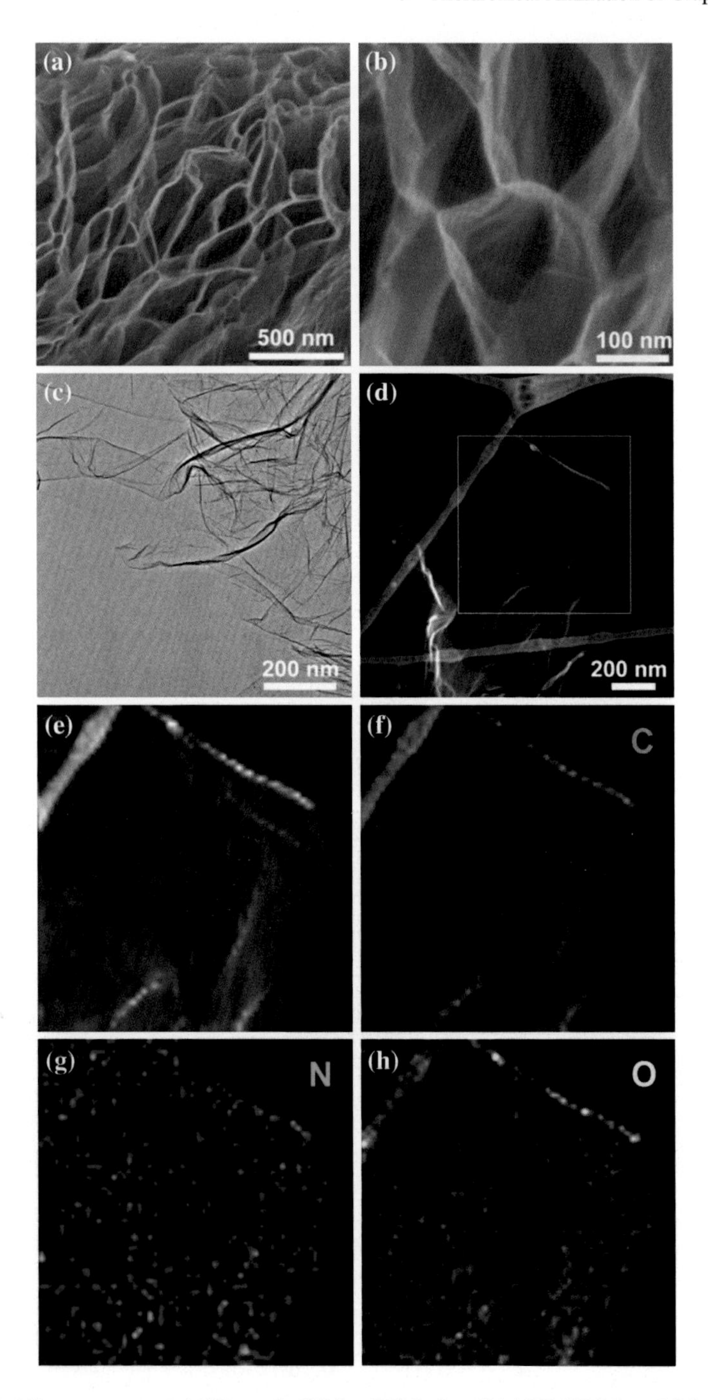

Fig. 3.1 Microstructure of AGHs: **a**, **b** SEM, **c** TEM, **d**, **e** HAADF-STEM, and **f**–**h** C, N, O elemental maps of AGH200. Reproduced from Ref. [63] by permission of The Royal Society of Chemistry

is composed of micrometer-sized curved flakes with a high-transparency chiffon-like texture. The HAADF image (Fig. 3.1e) and C, N, O elemental maps of a selected zone (Fig. 3.1f–h) illustrate a homogeneous distribution of N and O functionalities on graphene. On one hand, the exterior porous structure of 3D assembly facilitates fast ionic transportation and thus reduces diffusion resistance. On the other hand, uniform functionalization on graphene not only improves the hydrophilicity and wettability of the AGHs with the electrolyte but also participates in reversible redox reaction and generates pronounced pseudocapacitance.

3.3.2 The Surface Chemistry of AGHs

XPS analysis is carried out to obtain the evolution of surface chemistry from G250 to AGHs. The survey spectra and C1s fine scan spectra are shown in Fig. 3.2, while the relative content of C, N, and O elements on the surface of sample is calculated accordingly, as summarized in Table 3.1. As the amination temperature increases, oxygen atoms are gradually decreased to 5.23 at.% for AGH600 from 10.74 at.% for G250, with a simultaneous introduction of nitrogen (2.79 at.% for AGH200 to 3.91 at.% for AGH600). It is worth noting that AGH200 possesses the maximum heterogeneity (N + O ~13.09 at.%) among all the samples. As shown in Fig. 3.3a, b, various C (C1–C5) and O (O1–O5) components in AGHs are further determined by fitting the N1s and O1s fine scan spectra. The summary of XPS fitting results, which quantitate peak position and the relative surface concentration of nitrogen and oxygen species, is listed in Tables 3.2 and 3.3, respectively. Five types of nitrogen species can be observed on all sample surfaces: pyridinic, (N1 at 398.1 eV), amine/amide (N2 at

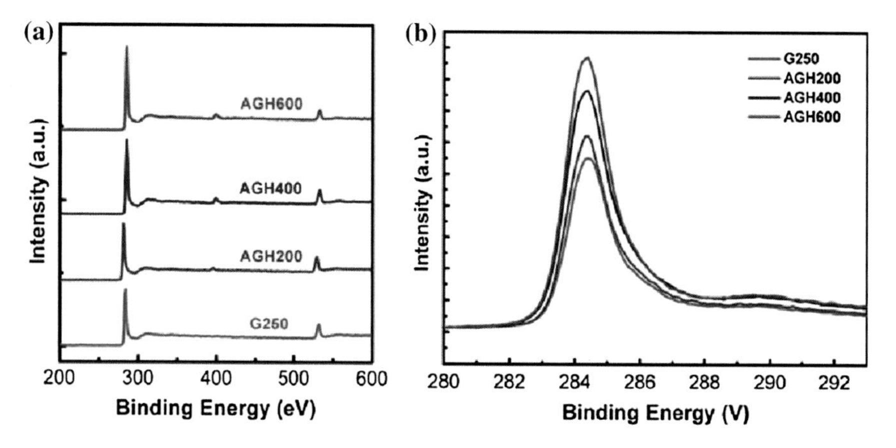

Fig. 3.2 **a** XPS survey spectrum and **b** C1s fine scan spectrum of G250 and AGHs. Reproduced from Ref. [63] by permission of The Royal Society of Chemistry

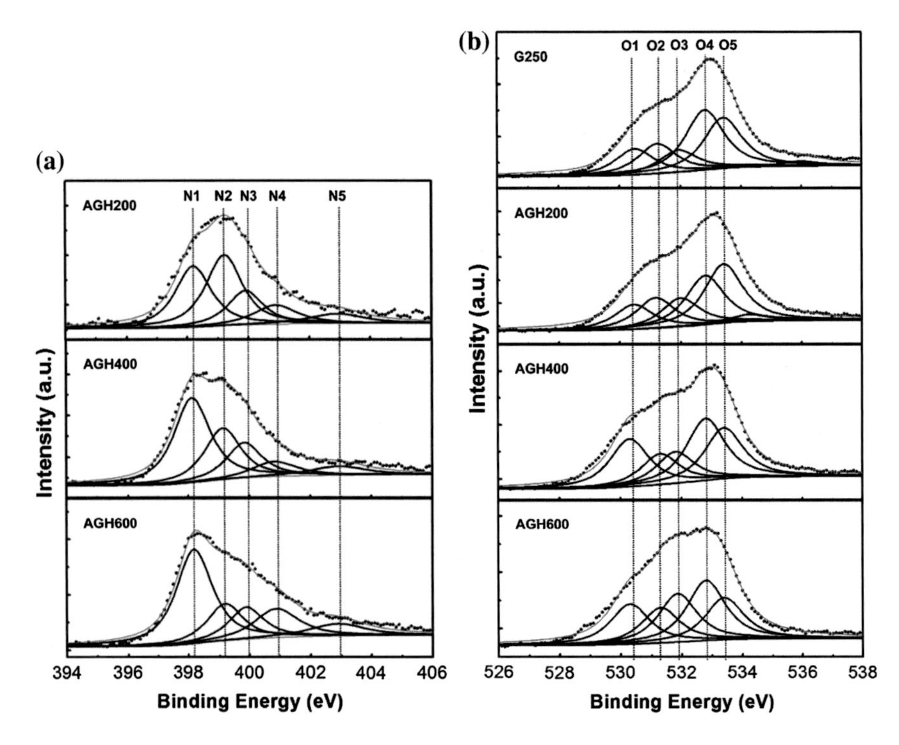

Fig. 3.3 XPS analysis of G250 and AGHs: Fitting results of **a** N1s and **b** O1s fine scan spectrum. Reproduced from Ref. [63] by permission of The Royal Society of Chemistry

399.3 eV), pyrrole/pyridine (N3 at 400.6 eV), quaternary nitrogen (N atoms incorporated in the graphitic layer in substitution of C atoms) (N4 at 402.1 eV), and N-oxide/nitro (N5 at 403.8 eV). While the O1s spectra of all the samples can be resolved into quinones (O1), COOH/C(O)O (O2), –C=O (O3), –C–O (O4), and –OH (O5), related groups peaks centered at 530.4 eV, 531.2 eV, 531.9 eV, 532.8 eV, and 533.4 eV, respectively [42].

As shown in Table 3.3, the pristine G250 contains abundant oxygen functional groups (–COOH, –OH, –C–O–C–, and –C=O), which has already been proved by previous reports [43, 44]. After amination at 200 °C, the primary N species in AGH200 has formed by the amine/amide-type N atoms chemically bonding to graphene (‘chemical nitrogen’). As the amination temperature increases, more pyridinic, pyrrolic, and quaternary type N atoms (‘lattice nitrogen’), which substitute carbon in the matrix, are increasing gradually with a simultaneous decrease of thermally unstable ‘chemical nitrogen’ [45]. Meanwhile, the acidic –COOH groups and –C–O related species are remarkably decreasing from G250 to AGH600 (from 1.63 to 0.81 at.% for –OOH, and from 3.31 to 1.39 at.% for –C–O), while –C=O and –OH species in AGH200 are

Table 3.2 Summary of nitrogen containing functional groups in AGHs

B.E. (eV)	N1 (398.1)	N2 (399.3)	N3 (400.6)	N4 (402.1)	N5 (403.8)
Species	Pyridinic	Amine/amide	Pyrrole/ pyridone	Quaternary	N-oxide/nitro
G250	0	0	0	0	0
AGH200	0.84	0.96	0.46	0.33	0.19
AGH400	1.57	0.94	0.65	0.34	0.24
AGH600	1.71	0.66	0.58	0.65	0.32

Reproduced from Ref. [63] by permission of The Royal Society of Chemistry

Table 3.3 Summary of oxygen containing functional groups in AGHs

B.E. (eV)	O1 (530.4)	O2 (531.2)	O3 (531.9)	O4 (532.8)	O5 (533.4)
Species	Quinones	COOH	C=O	C–O	OH
G250	1.42	1.63	1.22	3.31	2.78
AGH200	1.27	1.57	1.47	2.53	3.05
AGH400	1.80	1.15	1.19	2.34	1.92
AGH600	0.93	0.81	1.12	1.39	0.97

Reproduced from Ref. [63] by permission of The Royal Society of Chemistry

increased to a maximum ratio of 1.47 and 3.05 at.%, respectively. However, the two species will be gradually diminished in AGH400 and AGH600.

It can be seen from Fig. 3.2b that C1s intensity at 284.4 eV gradually increases from G250 to AGH600, which indicates the restoration of sp^2 conjugated basal plane of the graphene lattice due to the removal and/or stabilization of sp^3 saturated oxygen functionalities. Besides, O at.% decreases and N at.% increases simultaneity from G250 to AGH600, which suggests that a portion of oxygen sites and lattice defects on graphene are incorporated with or substituted by N-containing functional groups during amination, while another amounts of O species are temperately eliminated in the form of small molecules (e.g., H_2O, CO, and CO_2) [42, 44].

In agreement with the XPS results, N and O K edges in EELS spectroscopy illustrate a similar tendency (Fig. 3.4a). The carbon K-edge EELS spectra clearly present two energy loss characteristic peaks. The first narrow peak at around 284 eV corresponds to an electronic transition from carbon 1s to antibonding π^* states, which is the characteristic of sp^2-bonded carbon, while the broader structure at 290–310 eV indicates transitions to antibonding σ^* states. The distances between π^* and σ^* peaks for C (7.56 eV) and N (7.59 eV) species are almost the same, which reveal a homogeneous bulk doping of N atoms in AGHs.

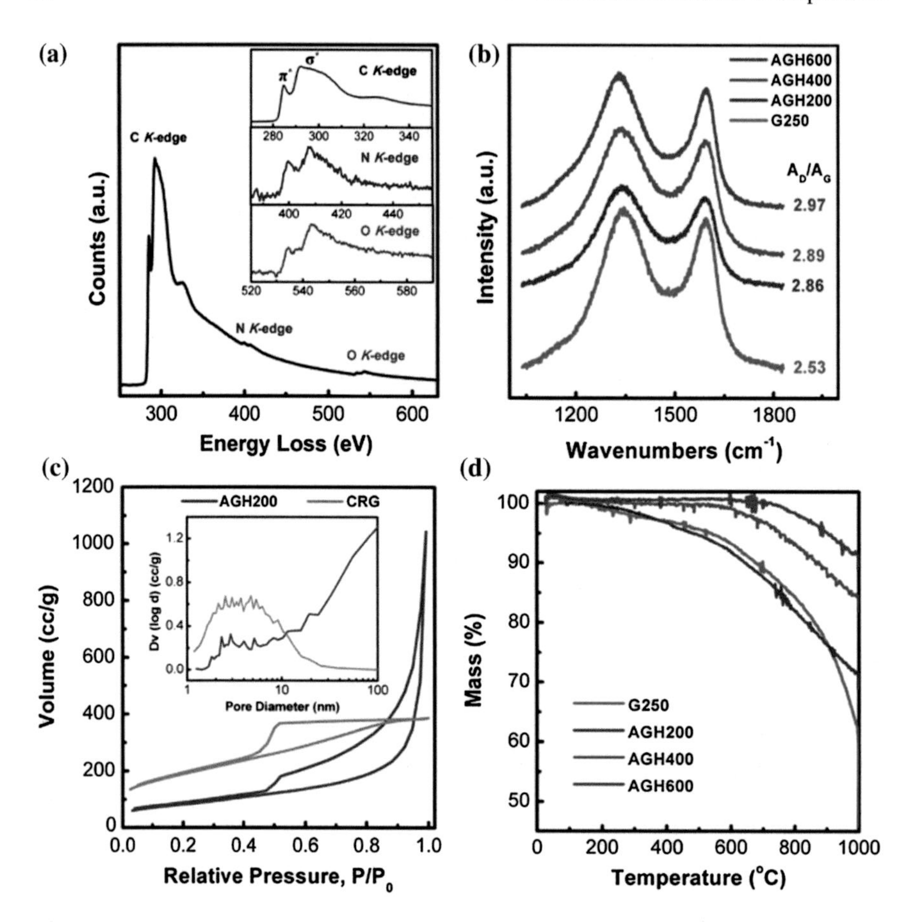

Fig. 3.4 Chemical analysis: **a** EELS spectra of AGH200, **b** Raman evolution from G250 to AGHs, **c** N_2 sorption isotherm and inset pore size distribution of AGH200 and CRG, **d** TGA curves of G250 and AGHs. Reproduced from Ref. [63] by permission of The Royal Society of Chemistry

The nature of graphitic and disorder in G250 and AGHs is further studied by Raman spectroscopy (Fig. 3.4b). Two strong peaks at around 1339 cm^{-1} (D band) and 1591 cm^{-1} (sp^2-bonded carbon; G band) are observed for all the samples, respectively. In detail, the D band is attributed to the disorder-induced band of graphitic carbon, while G band is assigned to the in-plane vibrations of graphite. As the amination temperature increases, the intensity ratio of the D and G bands (A_D/A_G) increases gradually from 2.53 of G250 to 2.97 of AGH600, owing to the amorphization of graphitic domains in graphene during amination. On one hand, the introduction of nitrogen, especially the 'lattice N' which is directly bonded to the in-plane sp^2 carbons, will cause the lattice distortion of graphitic region so as to transform into sp^3 domain. On the other hand, the simultaneous thermal evolution of oxygen such as ether-type groups will create various atomic vacancies and may even crack

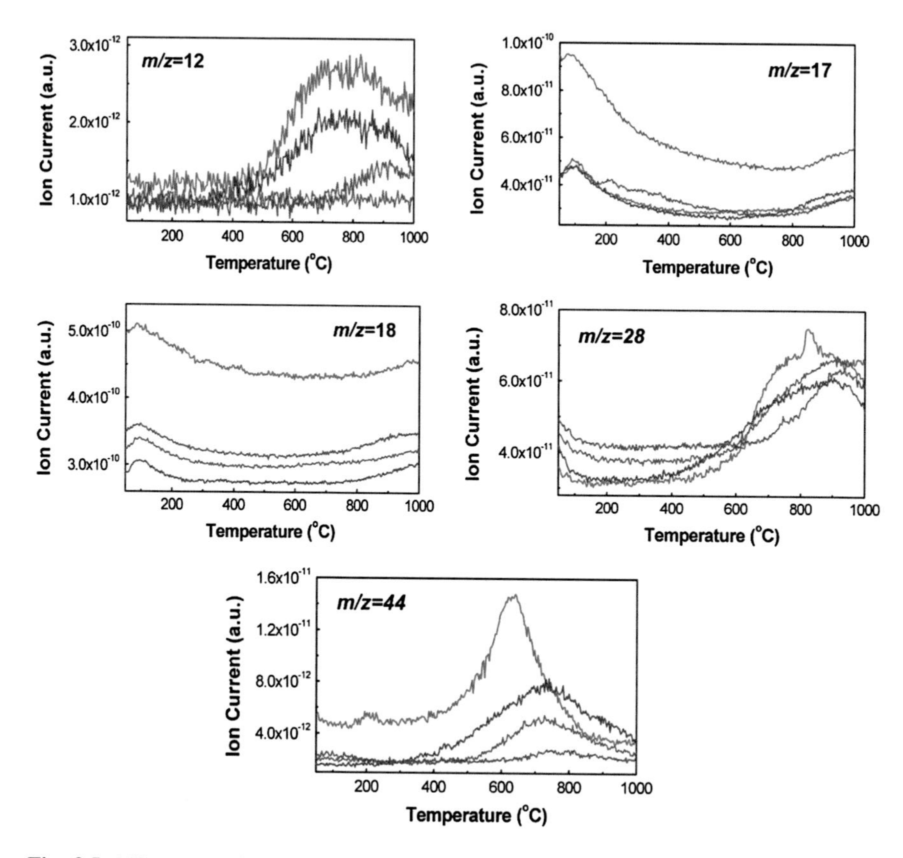

Fig. 3.5 MS curves of the sweep gases fragments from TG. Reproduced from Ref. [63] by permission of The Royal Society of Chemistry

the graphene sheets. Thereafter, numerous sp^2 graphitic domains with smaller sizes isolated by newly introduced amorphous regions are generated during the thermal amination process.

The N_2 physical adsorption is carried out to investigate the pore structure of AGH200 and CRG. As shown in Fig. 3.4c, the isotherms of AGH200 exhibit a type III adsorption isotherm with H3 hysteresis loop in IUPAC classification, which exhibits a unique adsorption behavior in the slit-shaped macropores raised from aggregates of plate-like particles. As shown in the BJH adsorption pore size distribution (inset of Fig. 3.4c), very little amount of pore volume of AGH200 is provided by micropores ($\Phi < 2$ nm). Moreover, the associated pore volume increases progressively with the increase of pore diameter, confirming the exterior macroporous structure of AGHs by SEM observation.

Because the great difference in thermal stability of various N and O groups, the surface chemistry of graphene (especially the relative abundance of various

Table 3.4 Assignments of TG–MS peaks

m/z	Assignments	Evolving temperature
12	Radical C evolved from re-graphitization or crystallization of amorphous sp^3 region in graphene	550–1000 °C
17	NH_3 from decomposition of nitrogen containing groups (especially for amine/amide species)	200–500 °C
	Radical OH fragments from H_2O (physically adsorbed water/chemical dehydration)	(a) ~100 °C (b) >800 °C
18	H_2O from evaporation of physically adsorbed water around 100 °C	~100 °C
	Intermolecular dehydration between neighboring –OH, –COOH and –NH_2 pairs	>800 °C
28	CO from decomposition of thermally stable –C=O related carbonyl groups, individual –OH or epoxy sites at relatively higher temperature	650–1000 °C
44	CO_2 from desorption of C(O)O related species such as carboxyl, lactone and anhydride	200 °C

Reproduced from Ref. [63] by permission of The Royal Society of Chemistry

functional groups) is largely dependent on the amination temperature employed [42]. Here, the thermogravimetric analysis–mass spectrometry (TG–MS) method is employed to analyze the thermochemistry of G250 and AGHs in Ar atmosphere, as shown in Figs. 3.4d and 3.5. The residual carbon of the materials after annealing to 1000 °C is summarized in Table 3.1. From G250 to AGH600, the residual carbon increases significantly. It is interesting that AGH200 (13.09 at.%) displays a higher hetero-atomic ratio than G250 (10.74 at.%), but the residual carbon of AGH200 (71.1 %) is higher than that of G250 (62.6 %). The sweep gases from TG are further analyzed by mass spectrometry, which can detect various fragments (e.g., CO_2, CO, NH_3, H_2O, and radical C) as shown in Table 3.4. After low-temperature amination, the amount of evolved carbon-containing gases (such as radical C, CO, and CO_2) decreases dramatically for AGH200, with a corresponding blue shift of the desorption temperatures for the functional groups. As the amination proceeds at relatively higher temperatures (e.g., 400 and 600 °C), most of the thermally unstable species (–COOH, –NH_2) are intensively eliminated or transformed into stable ones.

The structural and the accompanying surface chemical evolutions of N and O functional groups in different carbon allotropes (e.g., carbon black [46], activated carbon [47], carbon nanotubes [48], and graphene [44, 49, 50]) have been intensively studied in the past several decades. Figure 3.6 shows the possible types of nitrogen and oxygen species in aminated carbon materials. It is reported that the sequence of thermal stability for various oxygen containing groups on carbon surface is carbonyl > hydroxyl > epoxy > carboxyl [42, 44]. As shown in Fig. 3.7, in the initial amination stage (e.g., below 200 °C), NH_3 reacts with carboxylic acid species to form mainly intermediate amide or amine-like species ('chemical N') through

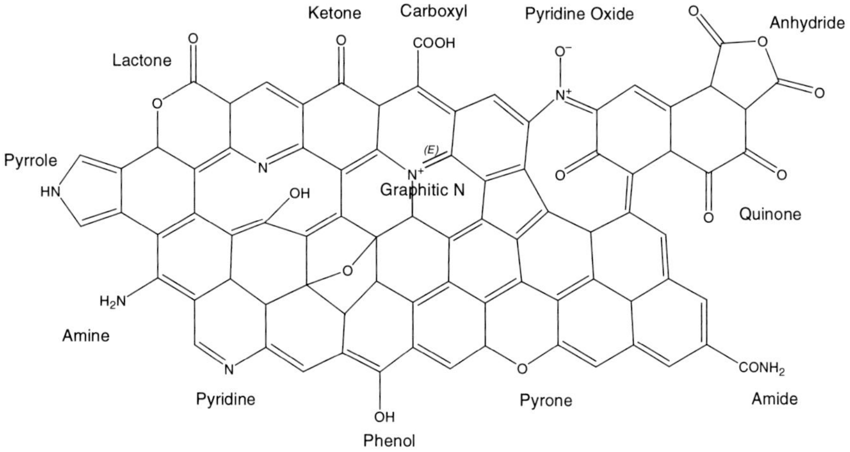

Fig. 3.6 Different forms of N and O functionalities in aminated carbon materials. Reproduced from Ref. [63] by permission of The Royal Society of Chemistry

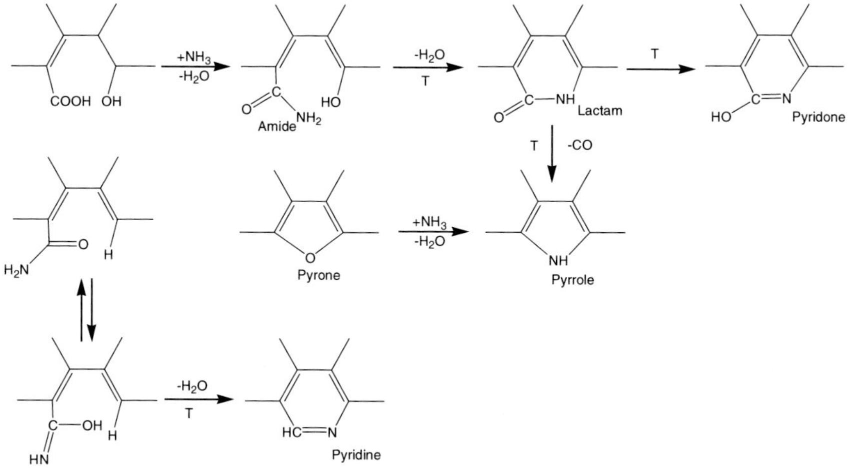

Fig. 3.7 Proposed pathways for N insertion into graphene. Reproduced from Ref. [63] by permission of The Royal Society of Chemistry

nucleophilic substitution. With the temperature increasing (e.g., 200–400 °C), more thermally stable heterocyclic aromatic moieties such as pyridine, pyrrole, and quaternary type N sites ('lattice N') have been formed through the intramolecular dehydration or decarbonylation.

3.3.3 The Electrochemical Capacitance of AGH

Figure 3.8a shows the CV curves of the G250 and AGHs at a slow scan rate of 3 mV s^{-1} in 6.0 M KOH aqueous electrolyte. Comparing with G250, AGH200 exhibits a significant enhancement of PC contribution in the potential range from −0.1 to 0.4 V, while AGH400 and AGH600 represent an increased ratio of EDLC character with a more rectangular shape in CV patterns. The capacitance values of G250, AGH200, AGH400, and AGH600 are given in Fig. 3.8b, c. It is noteworthy that the very high gravimetric capacitance C_F and specific capacitance C_s, 207 F g^{-1} and 0.84 F m^{-2}, respectively, are observed for AGH200, which is much higher than that of the others.

The enhancement of capacitive performance from G250 to AGH200 is mainly ascribed to the combined effect of the introduction of extra PC-active N species changing the electron donor/acceptor characteristics of AGH200 and the abundant original O containing functional groups preserved in AGH200 increasing the surface area accessibility for electrolyte ion transport [51]. This can further influence the electrochemical performance by affecting charging of the EDLC and to give extra PC [52].

As reported previously, the mechanism of O and N atoms obviously producing a beneficial effect on capacitance behavior in the aqueous electrolyte is mainly based on its faradaic reaction [40, 52]. Since the individual functional graphene sheet is consists of sp^2 conjugated aromatic domains decorated by abundant N and O species, the abundant delocalized electrons from the conjugated π bonds of graphitic domains and the extra lone pair electrons donated by the 'lattice N' supply a low-resistant pathway to charge transfer in the electrode. The inductive effects of the σ-bonded structure from O and N heteroatoms will cause a redistribution of the electrons as well as the polarization of some bonds. Thus, as depicted in Fig. 3.9, the reversible gaining/losing of electrons and simultaneous adsorption/desorption of protons will induce redox reactions of these polarized sites. Furthermore, the acidic functional species such as –COOH(II), –OH(III), and pyrrolic N(VII), which are well maintained or additionally introduced after amination at 200 °C, may play the key role in reversible Faradic redox reactions for generated pronounced pseudocapacitance in alkaline aqueous electrolyte (such as 6.0 M KOH employed in this work). Table 3.5 shows a comparison of the specific capacitances C_F (F g^{-1}) and C_s (F m^{-2}) of G250 and AGH200 with other carbon electrodes in 6 M KOH in the literature. The specific capacitances of the AGH200 electrode in this work are comparable to those previously reported for nanocarbon materials, which is ascribed to the efficient EDLC and PC contributions from each nanosheet within the hierarchical honeycomb architecture.

Apart from amination, the maximum temperature is employed for it effects in enhancing the electrochemical performance of graphene-based electrode. For example, the gravimetric capacitance of AGH400 (scan rate: 3 mV s^{-1}, $C_F = 104$ F g^{-1}) slumps to only ~50 % of AGH200, which result in a very poor specific capacitance (C_s) of only 0.15 F m^{-2}. On one hand, though more N atoms are introduced for AGH400 and AGH600, the total amount of heteroatom functionalities reduces remarkably due to an intensive removal of oxygen at high temperature, especially for the loss of some PC-active species (e.g., –C=O, –COOH and –OH). On the other hand, a higher treatment temperature of 400 or 600 °C will cause the collapse of hierarchical honeycomb-like structure (Fig. 3.10). Besides, though AGH400 and AGH600 have higher BET areas of 703 and 500 m^2 g^{-1}, respectively, the total pore volumes of both samples decreased and no micropore can be deduced from the *t*-plot analysis (Table 3.1). The above SEM and BET results reveal that more lattice defects are introduced to the basal plane of graphene and may even cause the crack of

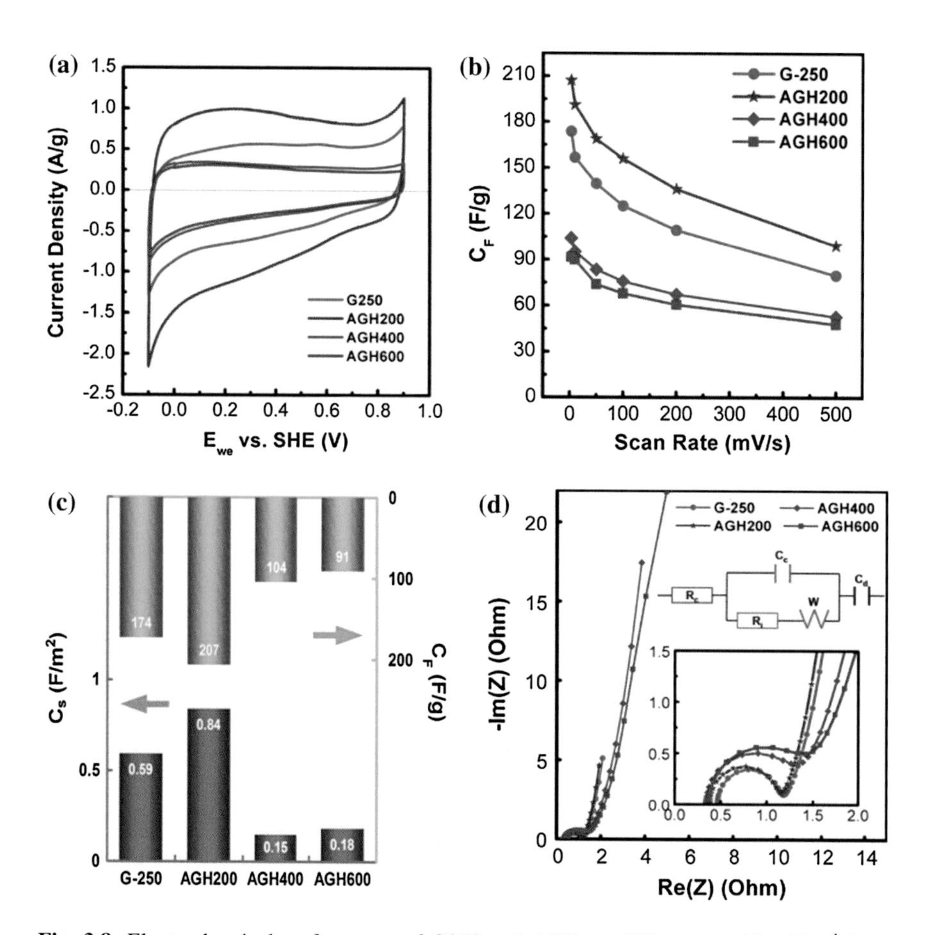

Fig. 3.8 Electrochemical performance of G250 and AGHs: **a** CV curves at 3 mV s^{-1} in an alkaline electrolyte (6.0 M KOH); **b** Gravimetric capacitance as a function of potential scan rates (from 3 to 500 mV s^{-1}); **c** Gravimetric and specific capacitance at 3 mV s^{-1}; **d** Nyquist plots of G250 and AGHs impedance spectra. Reproduced from Ref. [63] by permission of The Royal Society of Chemistry

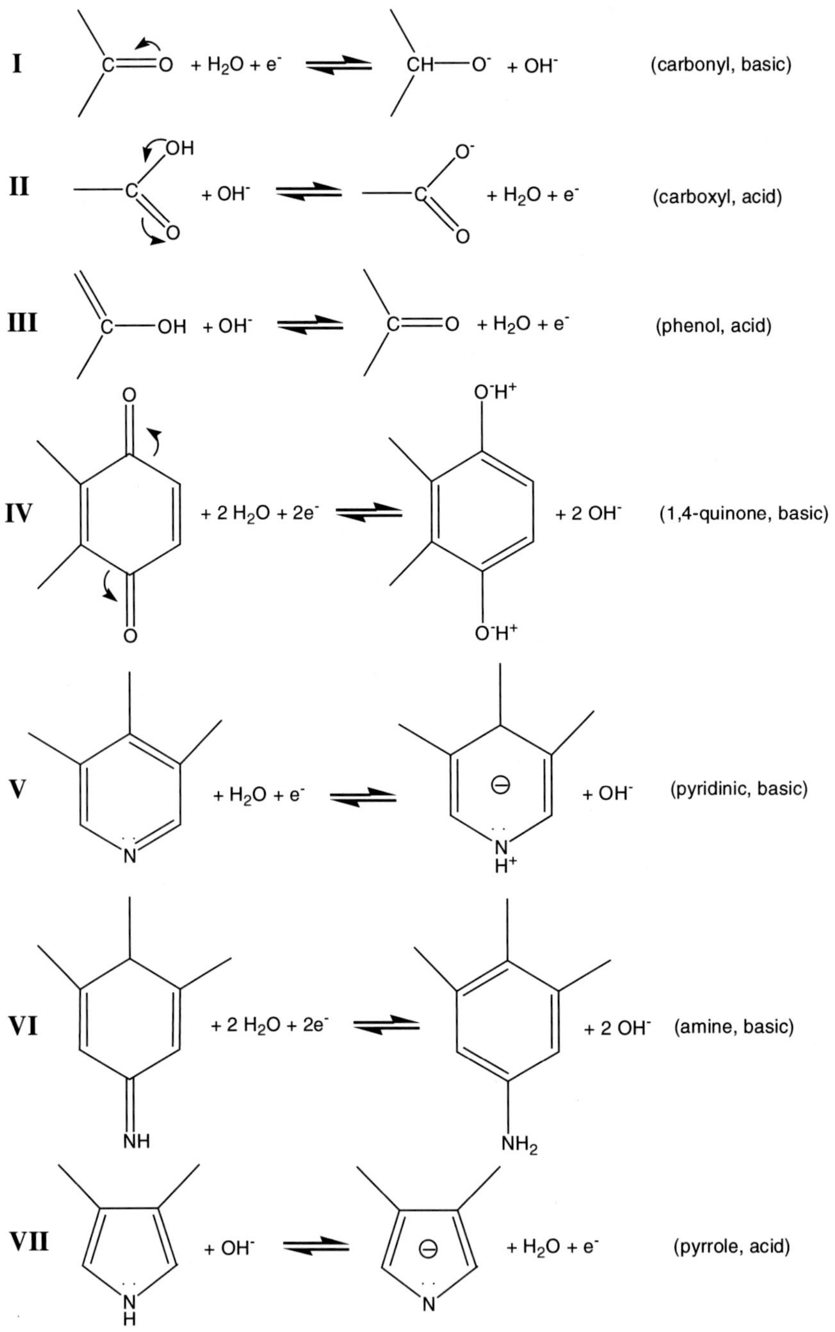

Fig. 3.9 Proposed reversible pseudocapacitive reactions of various functional sites. Reproduced from Ref. [63] by permission of The Royal Society of Chemistry

Table 3.5 Summary of supercapacitor performance based on carbon electrodes

Carbon type	BET surface area (m^2 g^{-1})	Scanning rate or cycling rate	C_F (F g^{-1})	C_s (F m^{-2})	Reference
G250	293	3 mV s^{-1}	174	0.59	This work
AGH200	247	3 mV s^{-1}	207	0.84	This work
Activated carbon	543	5 mV s^{-1}	184	0.34	[59]
LN[a]	746	0.2 A g^{-1}	264	0.35	[53]
LN + CNT	680	0.2 A g^{-1}	273	0.40	[53]
Activated carbon	1400	0.2 A g^{-1}	119	0.085	[53]
Mesoporous carbons	702	0.2 A g^{-1}	112	0.16	[64]
B0.7-OMC	641	0.2 A g^{-1}	134.6	0.21	[64]
P0.7-OMC	550	0.2 A g^{-1}	154	0.28	[64]
N-activated carbon	571	0.1 A g^{-1}	220	0.39	[58]
Graphene	202	10 mV s^{-1}	135	0.67	[4]
Graphene–CNT	612	10 mV s^{-1}	385	0.63	[4]
Graphene	705	20 mV s^{-1}	100	0.14	[17]
Graphene	382	1 mV s^{-1}	279	0.73	[65]
N-activated carbon	635	0.1 A g^{-1}	210	0.33	[66]
Carbide-derived carbons	600–2000	–	70–190	0.05–0.12	[67]
HOPG	–	–	–	0.5–0.7[b]	[68]
Graphite powder	4	–	1.4	0.35	[68]
Carbon aerogel	650	–	149.5	0.23	[68]
Nanodiamond	380	–	15.2	0.04	[69]
Carbon nanotube	200	–	18	0.09	[69]
Activated carbon	1150–2300	–	27–100	0.005–0.041	[40]
SWCNTs	357	1 mA cm^{-2}	138	0.39	[70]
MWCNTs	19.7	1 mA cm^{-2}	2	0.10	[71]
Activated MWCNTs	247	1 mA cm^{-2}	14	0.056	[71]

Reproduced from Ref. [63] by permission of The Royal Society of Chemistry
[a]Activated carbon obtained after pyrolysis at 600 °C of the Lessonia Nigrescens seaweed
[b]Measured at the edge plane of highly oriented pyrolytic graphite

which into smaller pieces during the harsh NH_3 activation process at higher temperature.

It is worth mentioning that the gravimetric capacitance of AGH200 electrode decreased from 207 to 98.95 F g^{-1}, with a very high capacitance retention ratio of 47.8 % at a high scan rate of 500 mV s^{-1}. This value is much higher than the conventional carbon electrode materials, such as activated carbon or mesoporous carbon (usually 20–30 %) [53, 54], which is ascribed to the tunable surface properties and the 3D porous structure of AGHs. First, the introduction of nitrogen as electron donor significantly enhances the charge carrier density in the basal

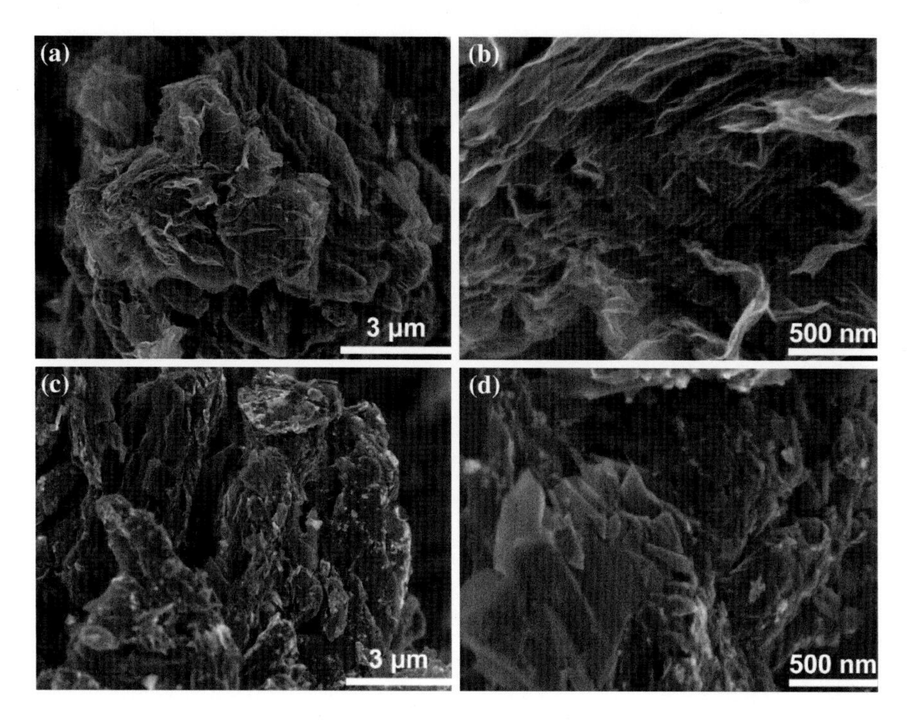

Fig. 3.10 Collapse of large pores in AGHs via high-temperature amination: SEM images of **a**, **b** AGH400 and **c**, **d** AGH600. Reproduced from Ref. [63] by permission of The Royal Society of Chemistry

plane [55–57], and the removal of some excessive oxygen groups effectively restored the sp^2 conjugated basal plane of graphene sheet. Thus, the conductivity of aminated graphene is correspondingly improved to render a facile electron transfer during the electrochemical charge–discharge process. Second, the existence of high contents of electrochemically active oxygen and nitrogen functional groups provides large amount of active sites for PC, due to their reversible redox reactions with ions in the electrolyte [27, 58]. Finally, the AGH features a 3D structure with macropores serving as ion-buffering reservoirs, and mesoporous walls significantly enhance electrolyte and electron transport with a shortened ion-diffusion distance as well as reduced resistance and large graphene surface providing more electroactive sites for energy storage charges [59, 60]. Therefore, the AGH materials show a promising prospect toward energy storage applications of fast charge–discharge capabilities.

To further confirm the double-layer formation, G250 and AGHs impedance spectroscopies are employed to distinguish the resistance and capacitance of the electrodes (Fig. 3.8d). An equivalent circuit model (inset of Fig. 3.8d) is introduced to simulate the capacitive and resistive elements of the cells under analysis [61, 62]. These elements include the internal resistance (R_i) of the AGH-based electrode, the

Table 3.6 Components of the equivalent circuit fitted for the impedance spectra

Sample	R_i (Ω)	R_c (Ω)	C_c (F/g)	W (Ω $S^{-1/2}$)	C_d (F/g)
G250	0.46	0.68	0.11	0.76	153.6
AGH200	0.38	0.73	0.24	0.66	192.2
AGH400	0.34	0.91	0.39	2.05	38.1
AGH600	0.37	0.99	0.54	2.49	36.2

capacitance, and resistance due to contact interface (C_c and R_c), a Warburg diffusion element attributable to the ion migration through the AGH (Z_w), and the capacitance inside the pores (C_d). The fitting results are shown in Table 3.6. In accordance with the microstructural and chemical evolution of graphene, the internal resistance (R_i) of AGH200 (0.38 Ω) is lower than the unaminated G250 (0.46 Ω). This is attributed to the effect of deoxygenation and nitrogen modification during amination, which not only restore the sp^2 conjugated carbon lattice, but also increase the charge carrier density. Furthermore, AGH200 exhibits a lower Warburg diffusion resistance ($Z_w = 0.66$ Ω $S^{1/2}$) than that of G250 ($Z_w = 0.76$ Ω $S^{1/2}$), which indicates improvement of ion transfer between inner channel of electrode and electrolyte. Compared with the unaminated G250 (C_d = 153.6 F g^{-1}), the inner-porous specific capacitance (C_d) of AGH200 could reach as high as 192.2 F g^{-1}, which is improved by ~25.1 % (Table 3.6).

To further confirm the contribution from the opened porous honeycomb structure in the electrode, a further control experiment with chemically reduced GO (named CRG) prepared by a typical liquid-phase reduction of GO by hydrazine hydrate is also performed. A disorderly packed morphology of agglomeration without hierarchical honeycomb structure has been observed for the CRG sample, as shown in Fig. 3.11a. In addition, it owns a high BET surface area of 666 m^2 g^{-1} and a considerable initial CF of 169 F g^{-1} at 3 mV s^{-1}.

However, the CF of CRG slumps to only 10.8 % at 500 mV s^{-1} (Fig. 3.11b). Figure 3.3c shows the N_2 sorption isotherm of CRG, which reveals the formation of ink-bottle pores, owing to the over-compact agglomeration of graphene sheets during chemical reduction and drying process. Besides, different from exterior macroporous structure of AGH200, CRG is mainly composed of the micro- and mesopores with a diameter of less than 11 nm, which provides the huge surface area (inset of Fig. 3.4c and Table 3.1). The 'necking' effect of these ink-bottle pores and internal worm-like channels will hamper the mass transfer and thus slow down the ion diffusion in CRG. The above result, from another perspective, further verifies the crucial role of 3D hierarchical honeycomb structure in improving the fast charge–discharge capabilities of supercapacitor electrodes during electrochemical cycling.

We calculated the power density and energy density from CV curves of a supercapacitor with a cell voltage window of 1.0 V [39]. Over a wide range of current drain time from 2 to 360 s, AGH200 owns an energy density of no less than 3.4 Wh kg^{-1}, with a corresponding power density as high as 6193 W kg^{-1}, which

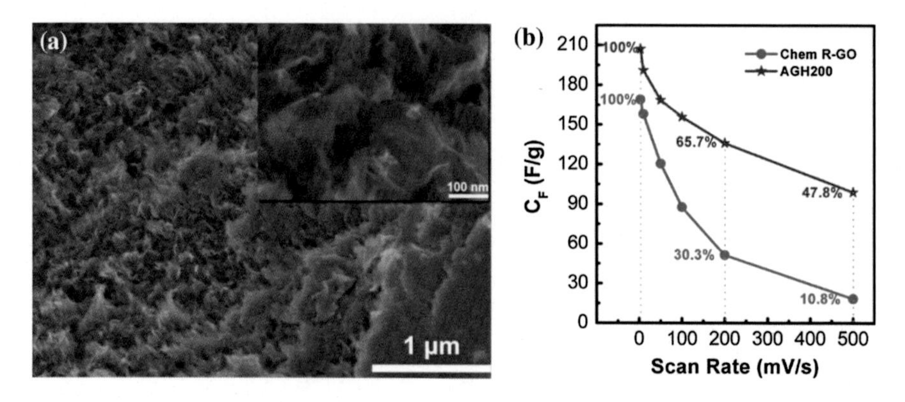

Fig. 3.11 Chemically reduced graphene oxide (CRG) by NH_2–NH_2: **a** SEM images of compact graphene agglomeration without exterior macropores; **b** Electrochemical performance as supercapacitor electrode. Reproduced from Ref. [63] by permission of The Royal Society of Chemistry

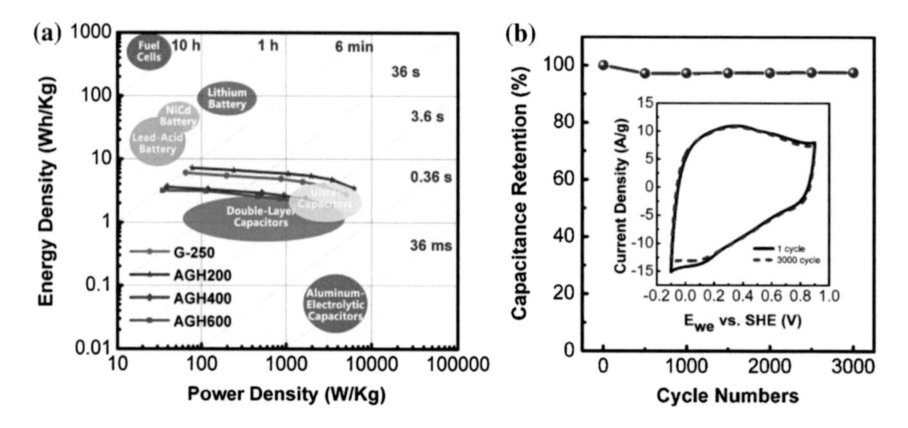

Fig. 3.12 Energy storage performance of AGHs: **a** Ragone plot showing the performance of graphene honeycomb based supercapacitor electrodes among various commercial energy storage devices (Source data from US Defense Logitics Agency); **b** Cycling performance of AGH200. *Inset* shows the 1 and 3000 cycle CV of AGH200 at a scan rate of 50 mV s^{-1} in 6.0 M KOH. Reproduced from Ref. [63] by permission of The Royal Society of Chemistry

is far beyond the normally used lithium ion batteries with a power density of less than 500 W kg^{-1}. At a long current drain time of 360 s, the energy density of AGH200 could reach as high as 7.2 Wh kg^{-1}, which is approaching that of commercial lead–acid batteries (Fig. 3.12).

In order to evaluate the cycling stability of the AGH electrode, repeating CV charge–discharge process was performed at a scan rate of 50 mV s^{-1} in the 6 M aqueous KOH electrolytes. After 3000 times of cycling, the gravimetric capacitance still maintain a high retention of 97.8 %, with only a tiny pattern change in CV curves, indicative of excellent long-term cycling durability. Compared with the other energy storage devices which often suffer from a short life cycle due to irreversible physical and/or chemical changes, AGH-based supercapacitor exhibits a prominent performance in durability and stability.

3.4 Conclusions

In this chapter, hierarchically aminated graphene honeycombs have been successfully prepared by the vacuum-assisted thermal expansion of graphite oxide and subsequent amination. When the low amination temperature (such as 200 °C) is employed, NH_3 reacts with carboxylic acid species to form mainly intermediate amide or amine-like species ('chemical N') through nucleophilic substitution. When the amination temperature increases, the intramolecular dehydration or decarbonylation will occur to form thermally stable heterocyclic aromatic moieties such as pyridine, pyrrole, and quaternary type N sites ('lattice N'). Because of PC enhancement by chemical modification of the 2D basic building block graphene and ion diffusion improvement through structural optimization of 3D honeycomb nanostructure, the functional 3D assembly of AGH200 shows excellent electrochemical performances with high specific capacitance and good cycling stability, as revealed by electrochemical measurements. In prospect, this indicates a promising way for hierarchical graphene honeycombs with tunable surface chemistry and mediated porous structure. The integration of graphene into hierarchical structures provides a maximum utilization of the excellent properties of graphene and could lead to promising 3D macrostructures. Such porous hierarchical architectures will benefit applications in heterogeneous catalysis, separation, and drug delivery, especially energy storage devices, such as Li-ion batteries, fuel cells, and supercapacitors, which require fast mass transfer through mesoporous, reactant reservoirs, and tunable surface chemistry.

References

1. Liu C, Li F, Ma LP, Cheng HM. Advanced materials for energy storage. Adv Mater. 2010;22(8):E28–62.
2. Hu H, Zhao ZB, Wan WB, Gogotsi Y, Qiu JS. Ultralight and highly compressible graphene aerogels. Adv Mater. 2013;25:2219–3.
3. Xu JJ, Wang K, Zu SZ, Han BH, Wei ZX. Hierarchical nanocomposites of polyanilinenanowire arrays on graphene oxide sheets with synergistic effect for energy storage. ACS Nano. 2010;4(9):5019–6.
4. Fan ZJ, Yan J, Zhi LJ, Zhang Q, Wei T, Feng J, et al. A three-dimensional carbonnanotube/graphene sandwich and its application as electrode in supercapacitors. Adv Mater. 2010;22(33):3723–8.
5. Yang SB, Cui GL, Pang SP, Cao Q, Kolb U, Feng XL, et al. Fabrication of cobalt and cobalt oxide/graphene composites: towards high-performance anode materials for lithium ion batteries. ChemSusChem. 2010;3(2):236–9.

6. Yu DS, Dai LM. Self-assembled graphene/carbon nanotube hybrid films for supercapacitors. J Phys Chem Lett. 2010;1(2):467–70.
7. Li SS, Luo YH, Lv W, Yu WJ, Wu SD, Hou PX, et al. Vertically aligned carbon nanotubes grown on graphene paper as electrodes in lithium-ion batteries and dye-sensitized solar cells. Adv Ener Mater. 2011;1(4):486–90.
8. Wu ZS, Wang DW, Ren W, Zhao J, Zhou G, Li F, et al. Anchoring hydrous RuO_2 on graphene sheets for high-performance electrochemical capacitors. Adv Funct Mater. 2010;20(20):3595–602.
9. Fan ZJ, Yan J, Wei T, Zhi LJ, Ning GQ, Li TY, et al. Asymmetric supercapacitors based on graphene/MnO_2 and activated carbon nanofiber electrodes with high power and energy density. Adv Funct Mater. 2011;21(12):2366–75.
10. Zhao MQ, Zhang Q, Jia XL, Huang JQ, Zhang YH, Wei F. Hierarchical composites of single/double-walled carbon nanotubes interlinked flakes from direct carbon deposition on layered double hydroxides. Adv Funct Mater. 2010;20(4):677–85.
11. Zhang Q, Zhao MQ, Liu Y, Cao AY, Qian WZ, Lu YF, et al. Energy-absorbing hybrid composites based on alternate carbon-nanotube and inorganic layers. Adv Mater. 2009;21(28): 2876–80.
12. Zhang LL, Zhao XS. Carbon-based materials as supercapacitor electrodes. Chem Soc Rev. 2009;38(9):2520–31.
13. Simon P, Gogotsi Y. Materials for electrochemical capacitors. Nat Mater. 2008;7(11):845–54.
14. Zhang H, Cao GP, Yang YS. Carbon nanotube arrays and their composites for electrochemical capacitors and lithium-ion batteries. Energy Environ Sci. 2009;2(9):932–43.
15. Lota G, Fic K, Frackowiak E. Carbon nanotubes and their composites in electrochemical applications. Energy Environ Sci. 2011;4(5):1592–605.
16. Novoselov KS, Geim AK, Morozov SV, Jiang D, Zhang Y, Dubonos SV, et al. Electric field effect in atomically thin carbon films. Science. 2004;306(5696):666–9.
17. Stoller MD, Park SJ, Zhu YW, An JH, Ruoff RS. Graphene-based ultracapacitors. Nano Lett. 2008;8(10):3498–502.
18. Bai H, Li C, Shi GQ. Functional composite materials based on chemically converted graphene. Adv Mater. 2011;23(9):1089–115.
19. Guo SJ, Dong SJ. Graphene nanosheet: synthesis, molecular engineering, thin film, hybrids, and energy and analytical applications. Chem Soc Rev. 2011;40(5):2644–72.
20. Zhu YW, Murali S, Stoller MD, Ganesh KJ, Cai WW, Ferreira PJ, et al. Carbon-based supercapacitors produced by activation of graphene. Science. 2011;332(6037):1537–41.
21. Sun YQ, Wu QO, Shi GQ. Graphene based new energy materials. Ener Environ Sci. 2011;4(4): 1113–32.
22. Loh KP, Bao QL, Ang PK, Yang JX. The chemistry of graphene. J Mater Chem. 2010;20(12): 2277–89.
23. Zhu YW, Murali S, Cai WW, Li XS, Suk JW, Potts JR, et al. Graphene and graphene oxide: Synthesis, properties, and applications. Adv Mater. 2010;22(35):3906–24.
24. Liu HT, Liu YQ, Zhu DB. Chemical doping of graphene. J Mater Chem. 2011;21(10): 3335–45.
25. Wang DW, Li F, Zhao JP, Ren WC, Chen ZG, Tan J, et al. Fabrication of graphene/polyaniline composite paper via in situ anodic electro polymerization for high-performance flexible electrode. ACS Nano. 2009;3(7):1745–52.
26. Yang SB, Feng XL, Wang L, Tang K, Maier J, Mullen K. Graphene-based nanosheets with a sandwich structure. Angew Chem Int Ed. 2010;49(28):4795–9.
27. Hulicova-Jurcakova D, Seredych M, Lu GQ, Bandosz TJ. Combined effect of nitrogen- and oxygen-containing functional groups of microporous activated carbon on its electrochemical performance in supercapacitors. Adv Funct Mater. 2009;19(3):438–47.
28. Jeong HM, Lee JW, Shin WH, Choi YJ, Shin HJ, Kang JK, et al. Nitrogen-doped graphene for high-performance ultracapacitors and the importance of nitrogen-doped sites at basal planes. Nano Lett. 2011;11(6):2472–7.

29. Dikin DA, Stankovich S, Zimney EJ, Piner RD, Dommett GHB, Evmenenko G, et al. Preparation and characterization of graphene oxide paper. Nature. 2007;448(7152):457–60.
30. Li XL, Zhang GY, Bai XD, Sun XM, Wang XR, Wang E, et al. Highly conducting graphene sheets and Langmuir-Blodgett films. Nat Nanotechnol. 2008;3(9):538–42.
31. Chen CM, Yang QH, Yang YG, Lv W, Wen YF, Hou PX, et al. Self-assembled free-standing graphite oxide membrane. Adv Mater. 2009;21(29):3007–11.
32. Kim F, Cote LJ, Huang JX. Graphene oxide: surface activity and two-dimensional assembly. Adv Mater. 2010;22(17):1954–8.
33. Park S, Mohanty N, Suk JW, Nagaraja A, An JH, Piner RD, et al. Biocompatible, robust free-standing paper composed of a TWEEN/graphene composite. Adv Mater. 2010;22(15):1736–40.
34. Yang XW, Zhu JW, Qiu L, Li D. Bioinspired effective prevention of restacking in multilayered graphene films: towards the next generation of high-performance supercapacitors. Adv Mater. 2011;23(25):2833–8.
35. Xu YX, Sheng KX, Li C, Shi GQ. Self-assembled graphene hydrogel via a one-step hydrothermal process. ACS Nano. 2010;4(7):4324–30.
36. Tang ZH, Shen SL, Zhuang J, Wang X. Noble-metal-promoted three-dimensional macroassembly of single-layered graphene oxide. Angew Chem Int Ed. 2010;49(27):4603–7.
37. Lee SH, Kim HW, Hwang JO, Lee WJ, Kwon J, Bielawski CW, et al. Three-dimensional self-assembly of graphene oxide platelets into mechanically flexible macroporous carbon films. Angew Chem Int Ed. 2010;49(52):10084–8.
38. Chen ZP, Ren WC, Gao LB, Liu BL, Pei SF, Cheng HM. Three-dimensional flexible and conductive interconnected graphene networks grown by chemical vapour deposition. Nat Mater. 2011;10(6):424–8.
39. Kotz R, Carlen M. Principles and applications of electrochemical capacitors. ElectrochimicaActa. 2000;45(15–16):2483–98.
40. Frackowiak E, Beguin F. Carbon materials for the electrochemical storage of energy incapacitors. Carbon. 2001;39(6):937–50.
41. Schniepp HC, Li JL, McAllister MJ, Sai H, Herrera-Alonso M, Adamson DH, et al. Functionalized single graphene sheets derived from splitting graphite oxide. J Phys Chem B. 2006;110(17):8535–9.
42. Figueiredo JL, Pereira MFR. The role of surface chemistry in catalysis with carbons. Catal Today. 2010;150(1–2):2–7.
43. Gao W, Alemany LB, Ci LJ, Ajayan PM. New insights into the structure and reduction of graphite oxide. Nat Chem. 2009;1(5):403–8.
44. Bagri A, Mattevi C, Acik M, Chabal YJ, Chhowalla M, Shenoy VB. Structural evolution during the reduction of chemically derived graphene oxide. Nat Chem. 2010;2(7):581–7.
45. Lota G, Lota K, Frackowiak E. Nanotubes based composites rich in nitrogen for supercapacitor application. Electrochem Commun. 2007;9(7):1828–32.
46. Boehm HP. Some aspects of the surface-chemistry of carbon-blacks and other carbons. Carbon. 1994;32(5):759–69.
47. Figueiredo JL, Pereira MFR, Freitas MMA, Orfao JJM. Modification of the surface chemistry of activated carbons. Carbon. 1999;37(9):1379–89.
48. Arrigo R, Havecker M, Wrabetz S, Blume R, Lerch M, McGregor J, et al. Tuning the acid/base properties of nanocarbons by functionalization via amination. J Am Chem Soc. 2010;132(28): 9616–30.
49. Wang XR, Li XL, Zhang L, Yoon Y, Weber PK, Wang HL, et al. N-doping of graphene through electrothermal reactions with ammonia. Science. 2009;324(5928):768–71.
50. Loh KP, Bao QL, Goki Eda, Chhowalla M. Graphene oxide as a chemically tunable platform for optical applications. Nat Chem. 2010;2(12):1015–24.
51. Xu B, Yue SF, Sui ZY, Zhang XT, Hou SS, Cao GP, et al. What is the choice for supercapacitors: graphene or graphene oxide? Ener Environ Sci. 2011;4(8):2826–30.
52. Frackowiak E. Carbon materials for supercapacitor application. Phys Chem Chem Phys. 2007;9(15):1774–85.

53. Raymundo-Piñero E, Cadek M, Wachtler M, Béguin F. Carbon nanotubes as nanotexturing agents for high power supercapacitors based on seaweed carbons. ChemSusChem. 2011;4(7): 943–7.
54. Zhu H, Wang XL, Yang F, Yang XR. Promising carbons for supercapacitors derived from fungi. Adv Mater. 2011;23(24):2745–8.
55. Wang XR, Li XL, Zhang L, Yoon Y, Weber PK, Wang HL, et al. N-doping of graphene through electrothermal reactions with ammonia. Science. 2009;324(5928):768–71.
56. Wei DC, Liu YQ, Wang Y, Zhang HL, Huang LP, Yu G. Synthesis of N-doped graphene by chemical vapor deposition and its electrical properties. Nano Lett. 2009;9(5):1752–8.
57. Lv RT, Cui TX, Jun MS, Zhang QA, Cao AY, Su DS, et al. Open-ended, N-doped carbon nanotube-graphene hybrid nanostructures as high-performance catalyst support. Adv Funct Mater. 2011;21(5):999–1006.
58. Zhao L, Fan LZ, Zhou MQ, Guan H, Qiao SY, Antonietti M, et al. Nitrogen-containing hydrothermal carbons with superior performance in supercapacitors. Adv Mater. 2010;22(45): 5202–5206.
59. Xu F, Cai RJ, Zeng QC, Zou C, Wu DC, Li F, et al. Fast ion transport and high capacitance of polystyrene-based hierarchical porous carbon electrode material for supercapacitors. J MaterChem. 2011;21(6):1970–6.
60. Wang DW, Li F, Liu M, Lu GQ, Cheng HM. 3D aperiodic hierarchical porous graphitic carbon material for high-rate electrochemical capacitive energy storage. Angew Chem Int Ed. 2008;47(2):373–6.
61. Huang CW, Hsu CH, Kuo PL, Hsieh CT, Teng HS. Mesoporous carbon spheres grafted with carbon nanofibers for high-rate electric double layer capacitors. Carbon. 2011;49(3):895–903.
62. Xu GH, Zheng C, Zhang Q, Huang JQ, Zhao MQ, Nie JQ, et al. Binder-free activated carbon/ carbon nanotube paper electrodes for use in supercapacitors. Nano Res. 2011;4(9):870–81.
63. Chen CM, Zhang Q, Zhao XC, Zhang BS, Kong QQ, Yang MG, et al. Hierarchically aminated graphene honeycombs for electrochemical capacitive energy storage. J Mater Chem. 2012;22:14076–84.
64. Zhao XC, Wang AQ, Yan JW, Sun GQ, Sun LX, Zhang T. Synthesis and electrochemical performance of heteroatom-Incorporated ordered mesoporous carbons. Chem Mater. 2010;22(19):5463–73.
65. Lv W, Tang DM, He YB, You CH, Shi ZQ, Chen XC, et al. Low-temperature exfoliated graphenes: vacuum-promoted exfoliation and electrochemical energy storage. ACS Nano. 2009;3(11):3730–6.
66. Yang XQ, Wu DC, Chen XM, Fu RW. Nitrogen-enriched nanocarbons with a 3-D continuous mesopore structure from polyacrylonitrile for supercapacitor application. J Phys Chem C. 2010;114(18):8581–6.
67. Chmiola J, Yushin G, Dash R, Gogotsi Y. Effect of pore size and surface area of carbide derived carbons on specific capacitance. J Power Sources. 2006;158(1):765–72.
68. Pandolfo AG, Hollenkamp AF. Carbon properties and their role in supercapacitors. J Power Sources. 2006;157(1):11–27.
69. Portet C, Yushin G, Gogotsi Y. Electrochemical performance of carbon onions, nanodiamonds, carbon black and multiwalled nanotubes in electrical double layer capacitors. Carbon. 2007;45(13):2511–8.
70. An KH, Kim WS, Park YS, Moon JM, Bae DJ, Lim SC, et al. Electrochemical properties of high-power supercapacitors using single-walled carbon nanotube electrodes. Adv Funct Mater. 2001;11(5):387–92.
71. Endo M, Kim YJ, Chino T, Shinya O, Matsuzawa Y, Suezaki H, et al. High-performance electric double-layer capacitors using mass-produced multi-walled carbon nanotube. Appl Phys A. 2006;82(4):559–65.

Chapter 4
Free-Standing Graphene Film with High Conductivity by Thermal Reduction of Self-assembled Graphene Oxide Film

4.1 Introduction

Graphene, as a two-dimensional crystal of sp^2 conjugated carbon atoms, is viewed as a building block for carbonaceous materials of other dimensionalities including zero-dimensional fullerenes, one-dimensional carbon nanotubes, and three-dimensional (3D) graphite [1]. It possesses a large surface area (2630 $m^2\ g^{-1}$ in theory), high electric charge carrier mobility ($\mu = 1.5 \times 10^4\ cm^2\ V^{-1}\ s^{-1}$), and excellent mechanical strength (tensile strength 130 GPa). However, aggregation or restacking inevitably occurs in graphene assemblies because of the intersheet van der Waals attractions, which causes many of the unique properties that individual sheets possess are significantly compromised or even unavailable in an assembly [2]. Thus, in order to fully verify the prominent properties of graphene, assembling 2D graphene sheets into macroscopic 3D architectures are highly concerned [3, 4]. Various 3D graphene architectures, such as graphene oxide (GO) [5–7] and/or graphene film [8–10], sponges [11], hydrogel [12], and foams [13–16], have been widely reported. As a consequence of their unique mechanical, electronic, and optical property, these materials have attracted intense research interest for their great potential for applications in batteries, supercapacitors, electronics, and sensors [7–10, 13–18]. For example, Ruoff and co-workers have developed GO films (GOFs) with a unique-layered structure and a tensile strength of 45–135 MPa, which can be easily fabricated by directional assembly of GO sheets. The as-obtained material exhibited attracting prospect for applications in films with controlled permeability, anisotropic ionic conductors, and supercapacitors [5]. Besides, macroscopic 3D graphene foams have also been fabricated by a template-directed chemical vapor deposition, which owned a unique network structure with outstanding electrical and mechanical properties [15]. Therefore, developing an effective and efficient method for free-standing graphene film is an important issue for the further exploration toward the property as well as the large-volume applications of graphene.

C.-M. Chen, *Surface Chemistry and Macroscopic Assembly of Graphene for Application in Energy Storage*, Springer Theses,
DOI 10.1007/978-3-662-48676-4_4

Among various physical and chemical synthesis approaches, oxidative intercalation and exfoliation of natural graphite and subsequent thermal or chemical reduction have been reported as an efficient strategic because of its low-cost, moderate product purity and rather large yield [2–4]. At the same time, the as-obtained chemically derived graphene sheets can be conveniently deposited on any substrate. So far, various solution processes such as filtration [5, 8–10, 19], spray coating [20], spin casting [21], Langmuir-Blodgett assembly [22], and self-assembly at the liquid/air interface [6] have been used to obtained large area GOFs. However, the as-obtained GOFs are almost insulating. Therefore, in order to restore the conductivity of GOFs, various deoxygenating approaches on GO powder or film have been developed, which can be roughly sorted as the 'dry' processes (high temperature annealing in vacuum, inert, or reductive atmosphere [20, 23–25]) and the 'wet' processes (chemical reduction by strong agents in solution, such as hydrazine [20, 26, 27], sodium borohydride [28], hydrohalic acid [29], aluminum powder [30], the sequential use of sodium borohydride, and sulfuric acid [31]). However, the GOFs, especially for the free-standing ones, are easily to be delaminated and thus cannot maintain the film-like morphology during the intensive reduction process. Therefore, to explore a simple method for the reduction of GOF is very important to maintain its free-standing film-like morphology as well as to achieve high electric conductivity. Thus, a free-standing GOF obtained by self-assembly at the liquid/air interface was annealed in a confined space between two stacked substrates to form a free-standing highly conductive graphene film.

4.2 Experimental

GO was prepared from graphite powder by the modified Hummers' method as previously reported [6, 32]. GO powder (3.0 g) was redispersed in 1.0 L of diluted water to obtain a stable hydrosol of GO with the concentration of ~2.8 mg mL^{-1} for further use. To prepare the GOF, the above hydrosol was heated at 80 °C for 40 min. Once the heating process was completed, a smooth and condensed thin film with the thickness of ca. 10 μm was collected from the liquid–air interface, and further air dried at 80 °C for another 8.0 h to form a free-standing film. Then, the film was laid between two stacked Si wafers to restore the conductivity of the GOF. In contrast, another control sample was placed freely without the confinement of an upper wafer. Both the two samples were put into a horizontal tubular furnace, and a heating schedule with a heating rate of 5 °C min^{-1} under the gas flow of Ar (80 mL min^{-1}) was executed, and then dwelled at 800 °C for another 2.0 h for further deoxidation. Afterward, the tube was cooled down to room temperature in Ar atmosphere. As a result, the film confined by two wafers during the annealing process maintained a very well free-standing morphology of flexible film, while the other one which was annealed freely without confinement finally collapsed into small fragments. The former obtained from the confined space was denoted as RGF, while the latter annealed freely without confinement was denoted as RGX.

The morphology of the as-obtained GOF, RGF, and RGX was characterized by JEOL JSM 7401F scanning electron microscope (SEM) operated at 3.0 kV; X-ray diffraction were obtained by X-ray diffractometer (XRD, Cu Kα radiation, D8 Advance, BRUKER/AXS, Germany); Fourier transform infrared spectroscopy (FT-IR) spectrums were collected on Thermo Nicolet IR200 spectrometer, and the sample was prepressed with KBr into pellets before test; X-ray photoelectron spectroscopy (XPS) was performed on Thermo VGESCALAB250 surface analysis system; the as-obtained graphene samples were further characterized using thermal gravimetric analysis (TGA) system (STA 409 PC Luxx, Netzsch, Germany) equipped with sweep gas mass spectroscopy appliance (ALZERS OmniStar 20, Switzerland). The electric conductivity of the film was performed on a Tongchuang SZT-2 Test Unit using four-point probe head with a pin-distance of about 1 mm.

4.3 Results and Discussion

As shown in Fig. 4.1a, a smooth and condensed thin film with a diameter of 8.5 cm and a thickness of 10 μm has been successfully obtained by the typical procedure. Figure 4.1b displays the cross-sectional SEM image of the GOF, which exhibits a compact layer-by-layer stacking structure. The top surface SEM images are shown in Fig. 4.1c and d indicate GO flakes served as 'building blocks' for the film. During the heating process, Brownian motion of GO sheets in the hydrosol is boosted. Thus, the GO sheets were prone to collide and interact with each other and move up to the liquid/air interface, where the water is spilling out from the hydrosol. The GO sheets tended to aggregate and assemble along the liquid/air interface. The nascent GO film will then capture the other GO sheets through interlayer van der Waals forces and began to stack. As a result, layer-by-layer film with ordered stacking morphology was formed [6]. In addition, ascribed to the poor conductivity of GOF (as shown in Table 4.1, $\sigma_s = 1.26 \times 10^{-5}$ S cm^{-1}), the charging behavior was observed during the SEM characterization.

After annealing of GOF confined by two stacked Si wafers in Ar atmosphere, a free-standing film was formed (Fig. 4.2a). The RGFs exhibit shining metallic luster because of the increase in reflectivity of visible light. Meanwhile, as determined by a 4-probe method, RGF shows a very high electrical conductivity with 272.3 S cm^{-1}, which is 7 magnitudes higher than that of raw GOF (Table 4.1). However, the film became fatigue and easily crushed into small debris. If the GOF was annealed in a free space without confinement by the upper substrate, the as-obtained film would crack itself into several small pieces (1–3 mm width and 5–15 mm in length, as shown in Fig. 4.2a) due to the random volume expansion in the annealing process. Furthermore, it can be seen from the cross-sectional SEM images (Fig. 4.2b and c) that the RGF maintains a regular morphology of layer-by-layer stacking of chemically derived graphene, and narrow pores (indicated by the blue arrow in Fig. 4.2b) were found in the RGF. The top surface of graphene sheets became crude (Fig. 4.2d–f), and the graphene sheets were crumpled with many wrinkles. No electron charging phenomena was observed, which was attributed to the good electronic conductivity of RGF.

The structural evolution during the annealing process was investigated by XRD, Raman Spectroscopy, FT-IR, XPS, and TG-DSC-MS. In contrast to the GOF film, the RGF film gives a detectable X-ray diffraction peak at 26.1° with a significant shrinking in lattice space (d_{002}) (from 0.702 to 0.341 nm) (Fig. 4.3a), which is attributed to the gradual dismissal of interlayer species such as physically adsorbed water and oxygen functionalities during annealing. The RGX exhibits a single peak around 26°; however, the diffraction intensity of which is quite lower than that of the confined annealed ones, which indicates a lower graphitization owing to random interlayer expansion during annealing. When the filtrated GOF was annealed at 150 °C, the interlayer space between reduced GO flakes (d_{002} value) was 0.441 nm [33]. With the annealing temperature increasing, the d_{002} value decreased. The packing of graphene sheets in RGF annealed at 800 °C were more compact than that of HI reduction (0.357 nm) [29], $NaBH_4$ reduction (0.373 nm) [28], HI-AcOH reduction (0.362 nm)

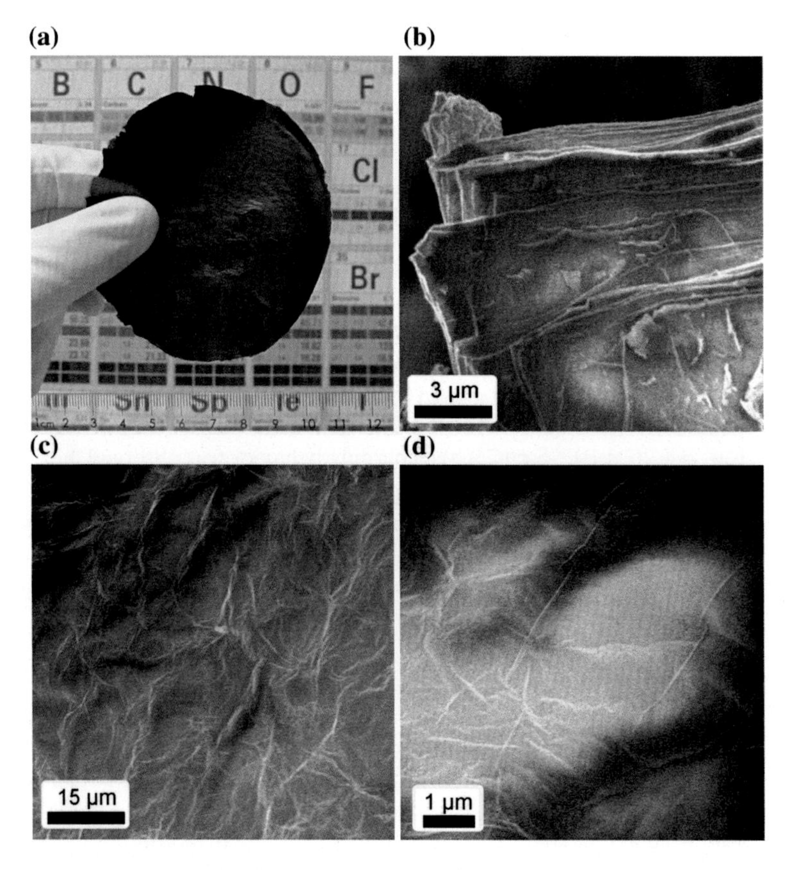

Fig. 4.1 **a** Photographs and (**b–d**) SEM images of flexible GOFs produced through a self-assembly process at the liquid/air interface. Reprinted from Ref. [43], Copyright 2012, with permission from Elsevier

Table 4.1 Summary of the structure and property of GOF and RGF

Sample	C (wt%)[a]	O (wt %)[a]	C/O atom ratio[a]	I_D/I_G[b]	Residue carbon (wt%)[c]	Conductivity (S/cm)[d]
GOF	51.6	46.2	1.49	1.02	44.2	1.26×10^{-5}
RGF	92.4	7.1	17.3	1.38	84.5	272.3

[a]Obtained from XPS
[b]Obtained from the Raman spectra
[c]Obtained from TGA in argon atmospheres
[d]Obtained by a four-probe method

[34], but relatively larger than that of graphite (0.337 nm) [35]. This indicated that most of the functional groups had been removed at high annealing temperature, which can be further verified by the fact that the 800 °C annealed RGF were with a high C/O ratio of 17.3. The structural evolution is further studied by Raman spectroscopy. The D-band around 1335 cm^{-1} is corresponding the defects or edges while the G-band around 1588 cm^{-1} is assigned to the first-order scattering of the E_{2g} mode [25], as presented in Fig. 4.3b. After annealing, the D-band down shifted to 1328 cm^{-1}, while the intensity ratio of D- to G-band increases progressively from 1.02 to 1.38. The I_D/I_G ratio of as-obtained film debris (RGX) by annealing GOF in a free space (1.42) is a little higher than that of RGF (1.38). During annealing of GO film, the loss of carbon atoms from the graphene oxide lattice results in the formation of defects such as vacancies and distortions, and this separated an integrate sp^2 domain into several smaller sp^2 crystalline, which results in the increase of the intensity ratio of D- to G-band in the RGF. Such phenomenon is similar to the reduction of GO by hydrazine hydrate [26, 36], $NaBH_4$ [28], or HI solution [29] as previously reported.

FT-IR spectrum was employed to confirm the loss of oxygen-containing groups after annealing. As shown in Fig. 4.4, the intensity of peak at 3700 cm^{-1} (−OH group) (identification similar to that reported in Refs. [9, 37]), 1682 cm^{-1} (−COOH group), 1533 cm^{-1} (−COO− group), and 1150 cm^{-1} (−CO− groups) was remarkably decreased. In order to further determine the evolution of functionalization in the film surface, XPS spectra of GOF and RGF were obtained, as shown in Fig. 4.5. The atomic percentage of C, O heteroatoms, as well as the C/O atomic ratio is calculated and summarized in Table 4.1. The C/O atom ratio was remarkably increased from 1.49 of GOF to 17.3 of RGF, which also confirms the removal of oxygen in the film during annealing. A further fitting of C1s and O1s fine scan spectra (Fig. 4.5) was conducted and the quantified results of different functional species were listed in Tables 4.2 and 4.3, respectively. The C1s spectra are composed of two main components arising from C–C (~284.4 eV) and C–O (epoxy and hydroxyl, ~286.4 eV), and two minor components from C=O (carbonyl, ~287.8 eV) and O–C = O (carboxyl, ~288.8 eV), which is in accordance with the previous reports [8, 38, 39]. Meanwhile, the total contents of oxygen functionalities decreased as the annealing temperature increased. The RGF annealed at 800 °C shows an obvious

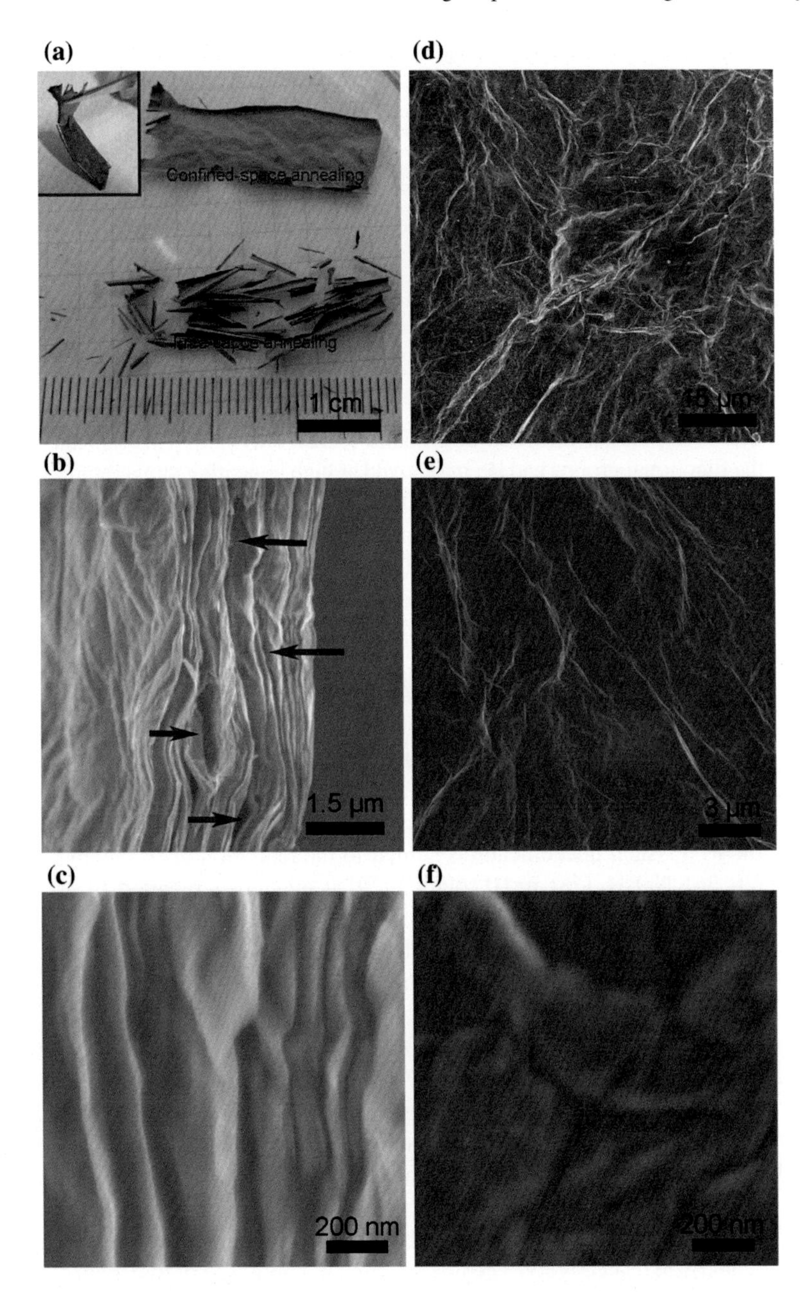

Fig. 4.2 **a** Photographs of RGF annealed between two stacked substrates and RGX on the *top* surface of a wafer. The RGF is flexible, illustrated by the inserted photo; (**b**–**f**) SEM images of free-standing RGF annealed between two stacked substrates. Reprinted from Ref. [43], Copyright 2012, with permission from Elsevier

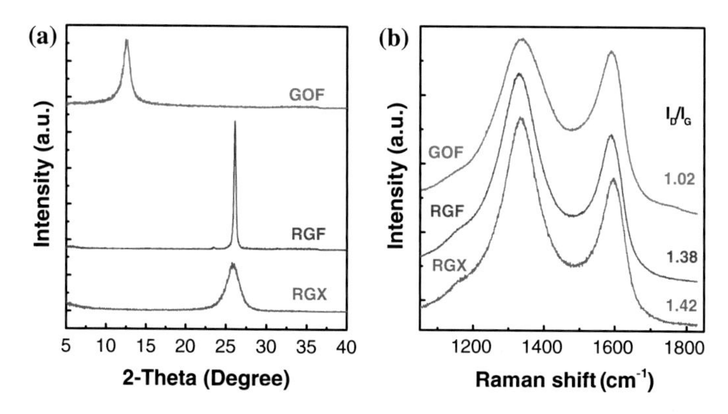

Fig. 4.3 **a** XRD patterns and **b** Raman spectra of GOF, RGF, and RGX. Reprinted from Ref. [43], Copyright 2012, with permission from Elsevier

decrease in oxygen components (such as C–O, C=O, O–H) with a simultaneous increase in graphitic carbon (such as C–C) (Table 4.2), and a graphitic shake up peak can thus be observed in the C1s spectra of RGF. The O1s spectra of GO consists of two major components from C–O (epoxy and hydroxyl, ~532.5 eV), C=O (carbonyl and carboxyl, ~531.6 eV), and two minor components for quinones (~530.5 eV) and O–H (hydroxyl, ~533.5 eV). After annealing, the normalized intensity of O1s peaks decreased sharply, especially for the contribution from C–O and C=O, indicating the removal of epoxy, carboxyl, and carbonyl functional groups. The contribution of O–H became the major one, and very few absorbed water at 535.1 eV can be detected on the RGF.

The thermochemistry of RGF is further analyzed by the TG-DSC analysis in both air and argon atmosphere (Fig. 2.5). As shown in Fig. 4.6, the RGF exhibits a prominent thermal stability with a very smooth weight loss starting from 800 °C in inert argon atmosphere and a residue carbon rate as high as 85 % at 1000 °C, which is favorable to supply RGF a fertile ground for further self-structural curing by carbonization and re-graphitization at a higher temperature (e.g., over 1200 °C) [40]. However, it remains an open question if the structural integrity of the films still preserved at these temperatures. Besides, the RGF also displays an excellent inflaming retarding performance and the kindling point could reach as high as ~630 °C in air, which is significantly higher than that of GOF (~450 °C as shown in Fig. 4.7).

TG-DSC-mass spectrometry (MS) experiment, by annealing GOF to 1000 °C in air and argon with in situ sweep gas analysis by MS, performed to examine, the decomposition behavior and chemical transformation of interlayer functional groups in GOF. As shown in Fig. 4.7a, the GOF presents a typical 3-stage weight loss pattern in inert atmosphere: (a) 13 % slowly weight loss between 0 and 170 °C, with a corresponding peak in MS at 95 °C ($m/z = 18.16$), which is expected to be the removal of the physically adsorbed interlayer water. The corresponding broad endothermic

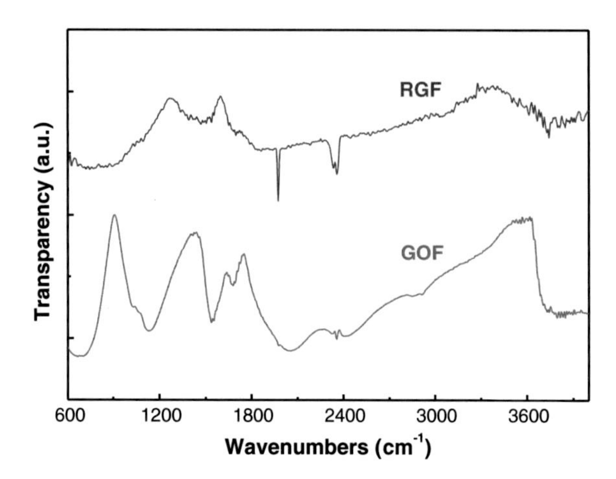

Fig. 4.4 FT-IR spectra of GOF and RGF. Reprinted from Ref. [43], Copyright 2012, with permission from Elsevier

peak around ~85 °C in DSC pattern indicates the heat adsorbing behavior during water evaporation; (b) 23 % rapid weight loss from 170 to 238 °C, with sharp MS peaks at ~220 °C for all the listed species in Fig. 4.7, which is believed that the decomposition of oxygen-containing functional groups takes place, and the released small molecules were determined to be H_2O ($m/z = 18.16$), CO ($m/z = 28.18$), CO_2 ($m/z = 44.09$) and radical C ($m/z = 12.30$). The sharp positive peak at 220 °C in DSC reveals the exothermic process because of the bond breaking of functionalities; (c) 19 % smooth weight loss from 238 to 1000 °C, with comparatively steady H_2O and CO_2, but increasing CO molecules releasing, indicating the further deoxygenation from the sp^2 graphitic lattice. Finally, when the temperature reaches 1000 °C, ~45 % of the mass is preserved.

Based on the above results, the chemical evolution of the film during annealing is schematically illustrated in Fig. 4.8. As the basic building block of GOF, the individual GO sheet is believed to contain two kinds of crystalline regions: (i) sp^2 regions composed of unoxidized graphitic carbon and (ii) sp^3 regions composed of aliphatic carbon due to the decoration and destruction of graphitic lattice by various oxygen-containing groups (–COOH, –OH, –C=O and epoxy), lattice defects, and dangling bonds. During the thermal annealing of GOF, the closely packed individual GO sheets, epoxy and hydroxyl groups were removed as CO, and carboxyl groups were removed as CO_2 [41] with the increasing of temperature under Ar atmosphere. During annealing process, carbonyl and ether groups were formed through transformation of the initial hydroxyl and epoxy groups, and the removal of carbon from the graphene plane was more likely to occur when the initial hydroxyl and epoxy groups were in close proximity to each other [42]. From the illustration showed in Fig. 4.8a, it is found that the loss of oxygen functional groups and carbon atoms caused shrink in the size of sp^2 domains, which was in agreement with the Raman

Table 4.2 Fitted results (%) of C1s XPS spectra of GOF and RGF

Sample	C–C	C–O	C=O	C(O)O
GOF	39.2	51.2	8.4	1.2
RGF	73.2	18.1	3.6	5.1

Reprinted from Ref. [43], Copyright 2012, with permission from Elsevier

Table 4.3 Fitted results (%) of O1s XPS spectra of GOF and RGF

Sample	Quinones	O=C	O–C	O–H
GOF	4.0	22.2	69.2	4.6
RGF	11.6	7.8	3.3	77.3

Reprinted from Ref. [43], Copyright 2012, with permission from Elsevier

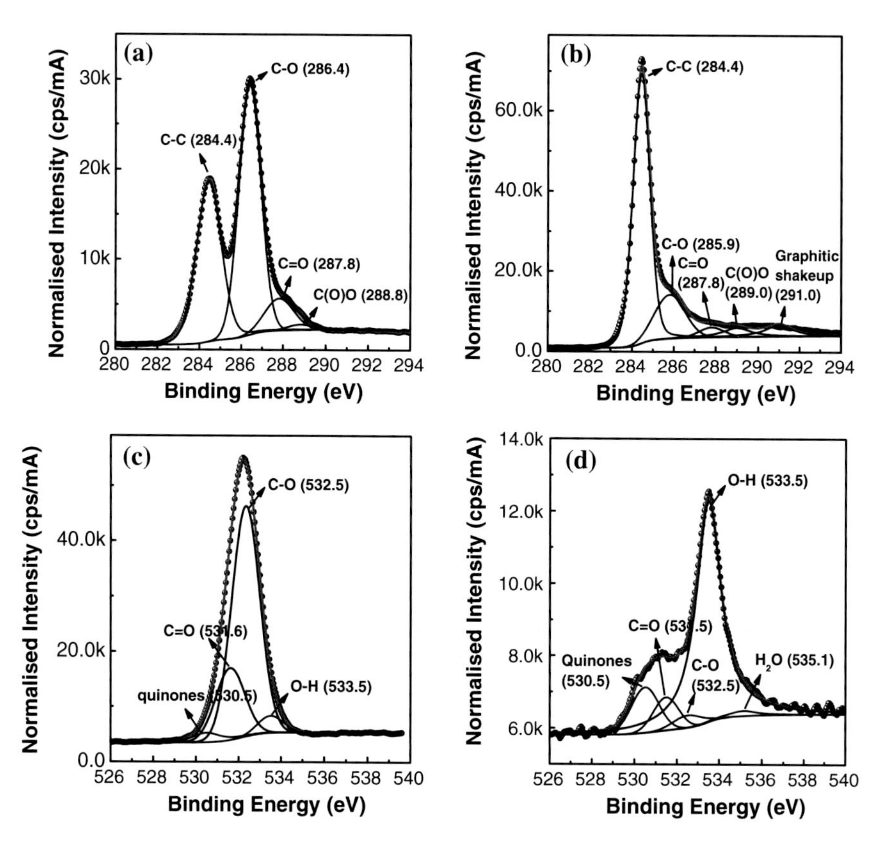

Fig. 4.5 XPS elemental fine scan analysis: C1s spectra of **a** GOF and **b** RGF, O1s spectra of **c** ROF and **d** RGF. Reprinted from Ref. [43], Copyright 2012, with permission from Elsevier

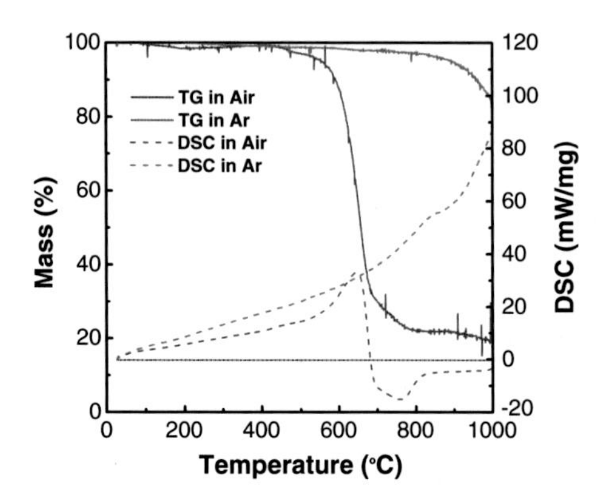

Fig. 4.6 TG-DSC analysis of RGF in air and argon atmospheres. Reprinted from Ref. [43], Copyright 2012, with permission from Elsevier

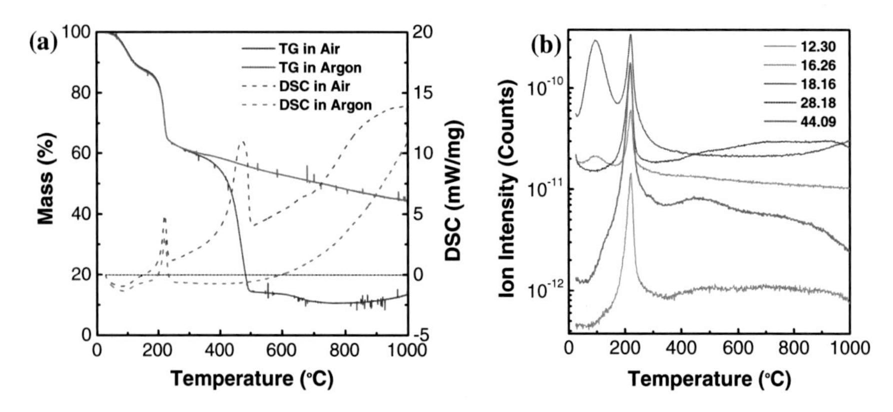

Fig. 4.7 **a** TG-DSC analysis of GOF in air and argon and **b** in situ MS spectra of sweep gas in the case of argon atmosphere. Reprinted from Ref. [43], Copyright 2012, with permission from Elsevier

spectra. On the other hand, the shrinks led to pristine internal stress with the loss of functional groups. If the annealing was taken place in a free space, cracks would generate between graphene sheets and the films would delaminate into twisted debris (Fig. 4.8b). However, if the film was confined between stacked wafers, the pressure-induced friction between the wafers and GO film would limit the movement of GO flakes, and thus the morphology could be well preserved (Fig. 4.8c). The improvement of electrical conductivity was highly dependent on the structural evolution during the annealing process. In GOF, during liquid-phase oxidation of graphite, the introduction of functionalities to both of the basal plane and edge of graphene caused a disruption of the C–C sp^2 conjugated crystal lattice, and resulted in the loss of conductivity. After annealing, reduced GO sheets and the increased π–π interactions between GO favored the transport of charge carriers.

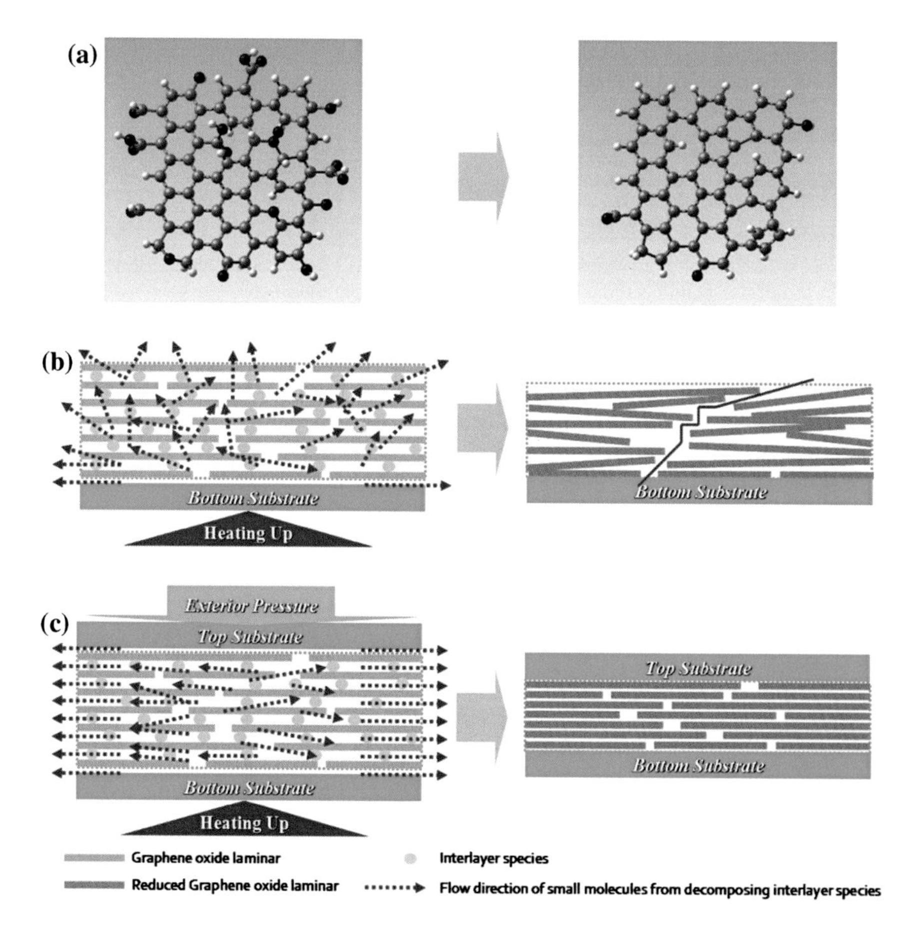

Fig. 4.8 Illustration re-graphitization of GOF into RGF through an annealing process: **a** Chemical evolution from GO to graphene; **b** Annealing of GOF in an open space; **c** Annealing of GOF in a confined space. Reprinted from Ref. [43], Copyright 2012, with permission from Elsevier

4.4 Conclusions

In summary, aggregation and restacking are the major hurdles that limit individual graphene sheets from realizing their full potential in an assembled bulk form. In this chapter, through annealing GO film between two stacked wafers, a highly conductive, free-standing graphene film was obtained. The oxygen-containing functional groups were removed during annealing, which caused the decrease of the sp^2 domains size. The conjugation of the graphene basal plane and π–π interactions between GO and reduced GO sheets favor the transport of charge carriers, and render the free-standing RGF with good electric conductivity. This is an efficient and effective way to fabricate highly conductive RGF in large scale and low cost, which will open up numerous opportunities for application exploration of graphene macroscopic materials.

References

1. Novoselov KS, Geim AK, Morozov SV, Jiang D, Zhang Y, Dubonos SV, et al. Electric field effect in atomically thin carbon films. Science. 2004;306(5696):666–9.
2. Stankovich S, Dikin DA, Piner RD, Kohlhaas KA, Kleinhammes A, Jia YY, et al. Synthesis of graphene-based nanosheets via chemical reduction of exfoliated graphite oxide. Carbon. 2007;45(7):1558–65.
3. Wei W, Yang SB, Zhou HX, Lieberwirth I, Feng XL, Müllen K. 3D graphene foams cross-linked with pre-encapsulated Fe_3O_4 nanospheres for enhanced lithium storage. Adv Mater. 2013;22:2909–14.
4. Gong YJ, Yang SB, Zhan L, Ma LL, Vajtai R, Ajayan PM. A bottom-up approach to build 3D architectures from nanosheets for superior lithium storage. Adv Funct Mater. 2014;24:125–30.
5. Dikin DA, Stankovich S, Zimney EJ, Piner RD, Dommett GHB, Evmenenko G, et al. Preparation and characterization of graphene oxide paper. Nature. 2007;448(7152):457–60.
6. Chen CM, Yang QH, Yang YG, Lv W, Wen YF, Hou PX, et al. Self-assembled free-standing graphite oxide membrane. Adv Mater. 2009;21(29):3007–11.
7. Kim F, Cote LJ, Huang JX. Graphene oxide: Surface activity and two-dimensional assembly. Adv Mater. 2010;22(17):1954–8.
8. Xu YX, Bai H, Lu GW, Li C, Shi GQ. Flexible graphene films via the filtration of water-soluble noncovalent functionalized graphene sheets. J Am Chem Soc 2008; 130(18): 5856.
9. An SJ, Zhu YW, Lee SH, Stoller MD, Emilsson T, Park S, et al. Thin film fabrication and simultaneous anodic reduction of deposited graphene oxide platelets by electrophoretic deposition. J Phys Chem Lett. 2010;1(8):1259–63.
10. Yang XW, Zhu JW, Qiu L, Li D. Bioinspired effective prevention of restacking in multilayered graphene films: Towards the next generation of high-performance supercapacitors. Adv Mater. 2011;23(25):2833–8.
11. Liu F, Seo TS. A controllable self-assembly method for large-scale synthesis of graphene sponges and free-standing graphene films. Adv Funct Mater. 2010;20(12):1930–6.
12. Xu YX, Sheng KX, Li C, Shi GQ. Self-assembled graphene hydrogel via a one-step hydrothermal process. ACS Nano. 2010;4(7):4324–30.
13. Lee SH, Kim HW, Hwang JO, Lee WJ, Kwon J, Bielawski CW, et al. Three-dimensional self-assembly of graphene oxide platelets into mechanically flexible macroporous carbon films. Angew Chem Int Ed. 2010;49(52):10084–8.
14. Tang ZH, Shen SL, Zhuang J, Wang X. Noble-metal-promoted three-dimensional macroassembly of single-layered graphene oxide. Angew Chem Int Ed. 2010;49(27):4603–7.
15. Chen ZP, Ren WC, Gao LB, Liu BL, Pei SF, Cheng HM. Three-dimensional flexible and conductive interconnected graphene networks grown by chemical vapor deposition. Nat Mater. 2011;10(6):424–8.
16. Liu QF, Ishibashi A, Fujigaya T, Mimura K, Gotou T, Uera K, et al. Formation of self-organized graphene honeycomb films on substrates. Carbon. 2011;49(11):3424–9.
17. Fan ZJ, Yan J, Zhi LJ, Zhang Q, Wei T, Feng J, et al. A three-dimensional carbon nanotube/graphene sandwich and its application as electrode in supercapacitors. Adv Mater. 2010;22(33): 3723–8.
18. Lv RT, Cui TX, Jun MS, Zhang Q, Cao AY, Su DS, et al. Open-ended, N-doped carbon nanotube-graphene hybrid nanostructures as high-performance catalyst support. Adv Funct Mater. 2011;21(5):999–1006.

19. Chen CM, Yang YG, Wen YF, Yang QH, Wang MZ. Preparation of ordered graphene-based conductive membrane. New Carbon Mater. 2008;23(4):345–50.
20. Pham VH, Cuong TV, Hur SH, Shin EW, Kim JS, Chung JS, et al. Fast and simple fabrication of a large transparent chemically-converted graphene film by spray-coating. Carbon. 2010;48(7):1945–51.
21. Xu YF, Long GK, Huang L, Huang Y, Wan XJ, Ma YF, et al. Polymer photovoltaic devices with transparent graphene electrodes produced by spin-casting. Carbon. 2010;48(11):3308–11.
22. Li XL, Zhang GY, Bai XD, Sun XM, Wang XR, Wang E, et al. Highly conducting graphene sheets and Langmuir-Blodgett films. Nat Nanotechnol. 2008;3(9):538–42.
23. Schniepp HC, Li JL, McAllister MJ, Sai H, Herrera-Alonso M, Adamson DH, et al. Functionalized single graphene sheets derived from splitting graphite oxide. J Phys Chem B. 2006;110(17):8535–9.
24. Eda G, Fanchini G, Chhowalla M. Large-area ultrathin films of reduced graphene oxide as a transparent and flexible electronic material. Nat Nanotechnol. 2008;3(5):270–4.
25. Yang D, Velamakanni A, Bozoklu G, Park S, Stoller M, Piner RD, et al. Chemical analysis of graphene oxide films after heat and chemical treatments by X-ray photoelectron and Micro-Raman spectroscopy. Carbon. 2009;47(1):145–52.
26. Stankovich S, Dikin DA, Piner RD, Kohlhaas KA, Kleinhammes A, Jia Y, et al. Synthesis of graphene-based nanosheets via chemical reduction of exfoliated graphite oxide. Carbon. 2007;45(7):1558–65.
27. Xu YX, Sheng KX, Li C, Shi GQ. Highly conductive chemically converted graphene prepared from mildly oxidized graphene oxide. J Mater Chem. 2011;21(20):7376–80.
28. Shin HJ, Kim KK, Benayad A, Yoon SM, Park HK, Jung IS, et al. Efficient reduction of graphite oxide by sodium borohydride and its effect on electrical conductance. Adv Funct Mater. 2009;19(12):1987–92.
29. Pei SF, Zhao JP, Du JH, Ren WC, Cheng HM. Direct reduction of graphene oxide films into highly conductive and flexible graphene films by hydrohalic acids. Carbon. 2010;48(15):4466–74.
30. Fan ZJ, Wang K, Wei T, Yan J, Song LP, Shao B. An environmentally friendly and efficient route for the reduction of graphene oxide by aluminum powder. Carbon. 2010;48(5):1686–9.
31. Gao W, Alemany LB, Ci LJ, Ajayan PM. New insights into the structure and reduction of graphite oxide. Nat Chem. 2009;1(5):403–8.
32. Hummers WS, Offeman RE. Preparation of graphitic oxide. J Am Chem Soc. 1958;80(6):1339.
33. Liu YZ, Li YF, Yang YG, Wen YF, Wang MZ. The effect of thermal treatment at low temperatures on graphene oxide films. New Carbon Mater. 2011;26(1):41–5.
34. Moon IK, Lee J, Ruoff RS, Lee H. Reduced graphene oxide by chemical graphitization. Nat Comm. 2010;1:73.
35. Liu J, Jeong H, Liu J, Lee K, Park JY, Ahn YH, et al. Reduction of functionalized graphite oxides by trioctylphosphine in non-polar organic solvents. Carbon. 2010;48(8):2282–9.
36. Liu JQ, Lin ZQ, Liu TJ, Yin ZY, Zhou XZ, Chen SF, et al. Multilayer stacked low-temperature-reduced graphene oxide films: Preparation, characterization, and application in polymer memory devices. Small. 2010;6(14):1536–42.
37. Lai LF, Chen LW, Zhan D, Sun L, Liu JP, Lim SH, et al. One-step synthesis of NH_2-graphene from in situ graphene-oxide reduction and its improved electrochemical properties. Carbon. 2011;49(10):3250–7.
38. Peng X-Y, Liu X-X, Diamond D, Lau KT. Synthesis of electrochemically-reduced graphene oxide film with controllable size and thickness and its use in supercapacitor. Carbon. 2011;49(11):3488–96.
39. Yan J, Wei T, Shao B, Ma FQ, Fan ZJ, Zhang ML, et al. Electrochemical properties of graphene nanosheet/carbon black composites as electrodes for supercapacitors. Carbon. 2010;48(6):1731–7.

40. Oberlin A. Carbonization and graphitization. Carbon. 1984;22(6):521–41.
41. Figueiredo JL, Pereira MFR, Freitas MMA, Orfao JJM. Modification of the surface chemistry of activated carbons. Carbon. 1999;37(9):1379–89.
42. Bagri A, Mattevi C, Acik M, Chabal YJ, Chhowalla M, Shenoy VB. Structural evolution during the reduction of chemically derived graphene oxide. Nat Chem. 2010;2(7):581–7.
43. Chen CM, Huang JQ, Zhang Q, Gong WZ, Yang QH, Wang MZ, et al. Annealing a graphene oxide film to produce a free standing high conductive graphene film. Carbon. 2012;50(2):659–67.

Chapter 5
Template-Directed Macroporous 'Bubble' Graphene Film for the Application in Supercapacitors

5.1 Introduction

Nanostructured carbon with assembled building blocks in diverse scales is of great importance for energy storage [1–3]. Graphene, as a two-dimensional crystal composed by sp^2 carbon atoms, has been assembled into three-dimensional (3D) macroscopic fibers [4], films [5, 6], and hybrids [7, 8], as well as porous materials [9–12] with multifunctional properties and applications.

Irreversible aggregation or restacking between graphene sheets tends to occur due to the huge interlayer van der Waals attractions. For instance, graphene-based paper-like materials, either fabricated by vacuum filtration [6], electrophoresis deposition [13], or self-assembly at the liquid-air [5, 14–16] and oil-water interface [17], always have a graphite-like stacking morphology, which becomes barrier for ion diffusion and inaccessible to the reactants. As a result, various strategies have been taken to prevent the restacking of graphene during the assembly process, such as (i) introduction of 'spacer' phases (e.g., carbon nanotubes [18–20], nanoparticles [21–25] and even 'water molecules' [26]) to form sandwich type structures and (ii) template-directed deposition or controlled assembly of graphene to obtain 3D porous structures (e.g., networks [12], foams [27], aerogels [10, 28–31], sponges [32, 33], and hydrogel [34]). For example, 3D graphene porous network was synthesized by directional chemical vapor deposition (CVD) on nickel foam, and the obtained materials showed application prospect in flexible energy storage [12]. In addition, as a cost-effective source of chemically derived graphene, graphene oxide (GO) hydrosol has been assembled into macroscopic porous materials by approaches, such as sol-gel cross-linking [10], hydrothermal synthesis [35], freeze drying [36], and thermolytic cracking [37].

The specially designed 3D porous structure efficiently prevent the restacking of graphene sheets, and shows low density, high porosity, strong mechanical toughness, large specific surface area, and/or high electron conductivity, etc. However, the porous material, which is assembled without externally induced orientation, renders

C.-M. Chen, *Surface Chemistry and Macroscopic Assembly of Graphene for Application in Energy Storage*, Springer Theses,
DOI 10.1007/978-3-662-48676-4_5

a random interconnection structure and exhibits isotropic macroproperties. In addition, the obtained 3D assemblies by the above methods are uneven in aperture, restricted by the applied template and/or technology. It is out of control.

The present chapter explored a novel hard template-directed ordered assembly for 3D macroporous bubble graphene film (MGF). It was supposed to be a green way for controllable and regular graphene film working as binder-free supercapacitor electrode with high rate capability, mechanical flexibility, and enhanced gravimetric capacitance.

5.2 Experimental

5.2.1 *Synthesis of PMMA Latex Spheres*

Monodisperse PMMA latex spheres were prepared by the emulsifier-free emulsion polymerization. Polymerization of methyl methacrylate (MMA) monomer was carried out in a three-necked bottom flask using potassium persulfate ($K_2S_2O_8$, KPS) as the initiator. Typically, 1000 mL deionized water was stirred at 350 rpm at 75 °C and purged with Ar. After stabilization of the temperature at an elevated level, 400 mL methyl methacrylate and 10 g KPS were added subsequently and the reaction was allowed to proceed for 2 h, producing colloidal PMMA latex spheres. PMMA latex spheres were harvested by centrifugation, and washed with deionized water once and then with ethanol twice. The as-prepared PMMA spheres were re-dispersed in deionized water to obtain an aqueous suspension (200 mg mL^{-1}) for further application.

5.2.2 *Hard Template Route for Macroporous Graphene Film*

Graphite oxide (GO) was prepared by a modified Hummers' method, followed by ultrasonication (100 W, 45 min) in deionized water to get the graphene oxide hydrosol (3 mg mL^{-1}). The graphene oxide hydrosol (20 mL) was mixed with PMMA sphere suspension (1 mL) and sonicated for 30 min to obtain a homogeneous colloidal suspension, which was then vacuum filtrated on a Millipore filter to obtain the sandwich type assembly of PMMA sphere and graphene oxide sheets. The above composite film was peeled off from the filter and air dried at 80 °C overnight, and the weight ratio of graphene oxide to PMMA spheres in the film was deduced to be 3:10. The composite membranes were annealed to 800 °C (at the heating rate of 5 °C min^{-1}) and dwelled for 30 min in a tubular furnace. The graphene oxide within the membrane was thermally reduced into graphene, while the PMMA spheres templates were removed simultaneously, so as to get the MGFs. A control sample was prepared by a similar procedure but without PMMA template, resulting in a compact graphene film (CGF).

5.2.3 *Structural Characterization*

The morphology of the samples was characterized by Hitachi S4800 scanning electron microscope (SEM) operated at 2.0 kV and Philips CM200 LaB6 transmission electron microscope (TEM) operated at 200.0 kV. Nitrogen adsorption–desorption measurements were performed at −196 °C with Micromeritics ASAP2010 instrument. Prior to the measurements, the samples were degassed at 110 °C for 1 h and at 250 °C for 3 h. The specific surface areas were calculated with BET equation and the average pore diameters were estimated with desorption branches based on BJH model. The obtained GO, PMMA, and GO/PMMA samples were further characterized using thermal gravimetric analysis (TGA) system (STA 409PC luxx, Netzsch, Germany).

5.2.4 *Electrochemical Measurements*

The electrochemical properties of MGF and CGF were measured in aqueous system (electrolyte: 6.0 M KOH). A three-electrode system was employed in the measurements, among which MGF/CGF served as the working electrode, platinum foil electrode as counter electrode, and saturated hydrogen electrode (SHE) as reference electrode. Cyclic voltammogram (CV) curves (scan rates varying from 3 to 1000 mV s^{-1}), Galvanostatic charge-discharge (GC) measurement, and electrochemical impedance spectroscopy (EIS) profiles were measured with VSP BioLogic electrochemistry workstation. The electrochemical capacitances were obtained both from CV curves. The Nyquist plot was fitted by EC-Lab software. The specific capacitance of the single electrode (C_{elec}) was calculated from CV curves based on the equation: $C_{elec} = i\mathrm{d}t/\mathrm{d}v$, where i is the instant current in Amperes per gram of active material (A g^{-1}), and $\mathrm{d}v/\mathrm{d}t$ is the scanning rate in Volts per second (V s^{-1}).

The specific capacitance of the cell (C_{cell}) was calculated from GC profiles based on the following equation: $C_{cell} = (I\Delta t)/\Delta U$, where I is the discharge current in Amperes per gram of active material (A g^{-1}), Δt is the discharge time in second (s), and ΔU is the voltage window from the end of the IR_{drop} to the end of the discharge process in Volts (V).

5.3 Results and Discussion

As shown in Fig. 5.1a, the mixture of GO hydrosol and PMMA spheres suspension was filtrated with the assist of vacuum to prepare the sandwich type assembly of PMMA spheres and GO sheets. The monodisperse polymethyl methacrylate (PMMA) latex spheres were used as the hard templates. The composite film was air dried and heated at 800 °C to remove the PMMA template and thermally reduce GO into graphene simultaneously. The gray free-standing PMMA/GO film (Fig. 5.1b) turned black after calcination, while the morphology of flexible graphene film was preserved well (Fig. 5.1c).

Scanning electron microscopy (SEM) and transmission electron microscopy (TEM) were adopted to observe the microstructures of composite film. As shown in Fig. 5.2a, the PMMA spheres were homogeneously covered with wrinkled GO sheets for the strong interaction of hydrogen bonding between PMMA sphere and GO sheets. Figure 5.2b exhibits the colloid quasi-crystalline structures of the packed uniform spheres, among the interface of which GO nanosheets were closely distributed. After thermal treatment, the lamellar structure of the GO/PMMA, contributing

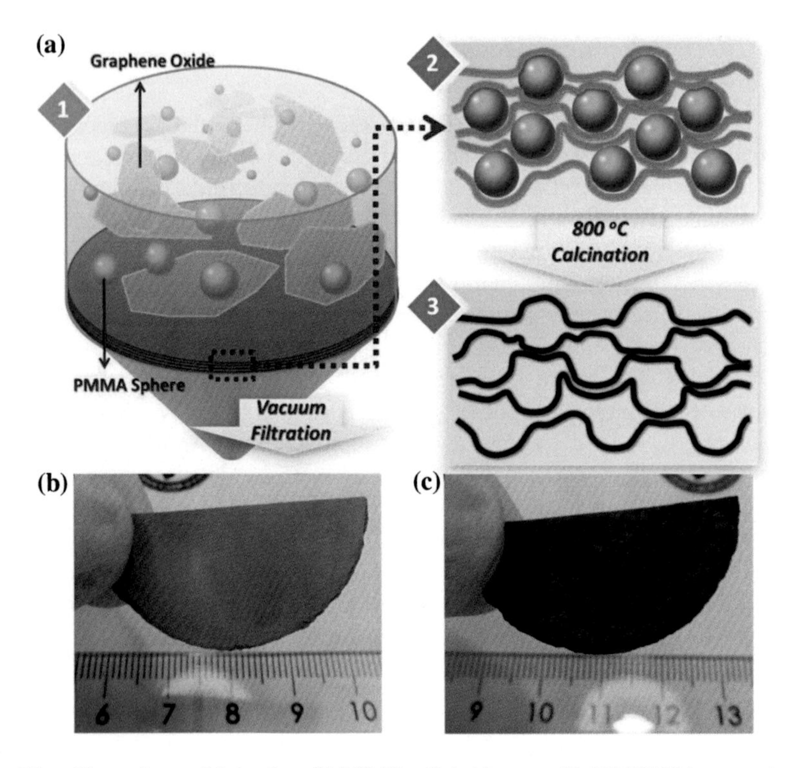

Fig. 5.1 **a** Illustration on fabrication of MGF. The digital images of **b** GO/PMMA composite film and **c** as-prepared MGF. Reproduced from Ref. [38] by permission of The Royal Society of Chemistry

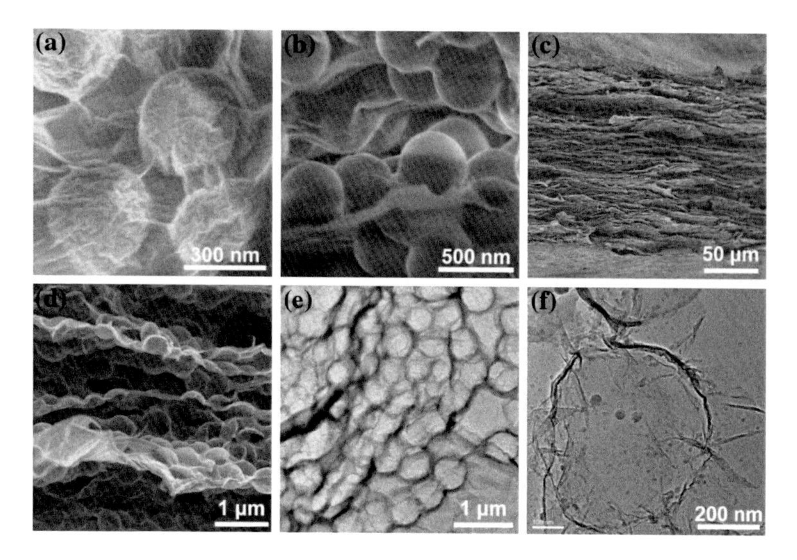

Fig. 5.2 SEM images of the **a** surface and **b** cross-section of a GO/PMMA composite film. **c** Low and **d** high magnification SEM images of the cross-section of a MGF. **e** Low and **f** high resolution TEM images of the graphene bubbles within a MGF sample. Reproduced from Ref. [38] by permission of The Royal Society of Chemistry

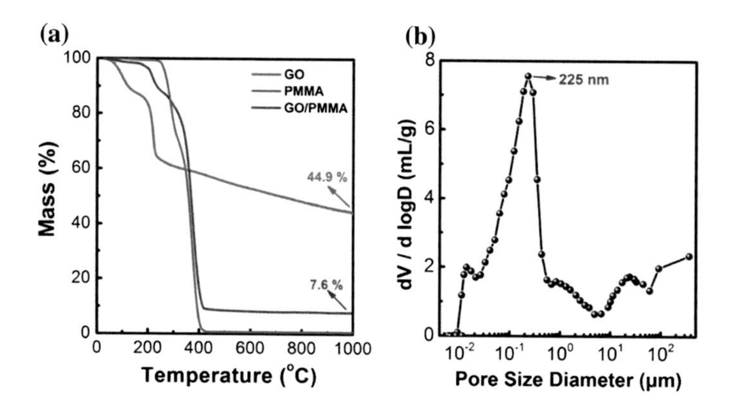

Fig. 5.3 **a** TG curves of GO, GO/PMMA composite and bare PMMA template. **b** Pore size distribution of MGF by Hg penetration analysis. Reproduced from Ref. [38] by permission of The Royal Society of Chemistry

a lot to the great mechanical strength of the integral free-standing film, was well-preserved (Fig. 5.2c). Meanwhile, the PMMA was pyrolyzed to gaseous products, with open hemispherical graphene nanosheets left in the MGF (Fig. 5.2d). The TEM images (Fig. 5.2e, f) further confirmed the hollow macroporous bubbles to be the

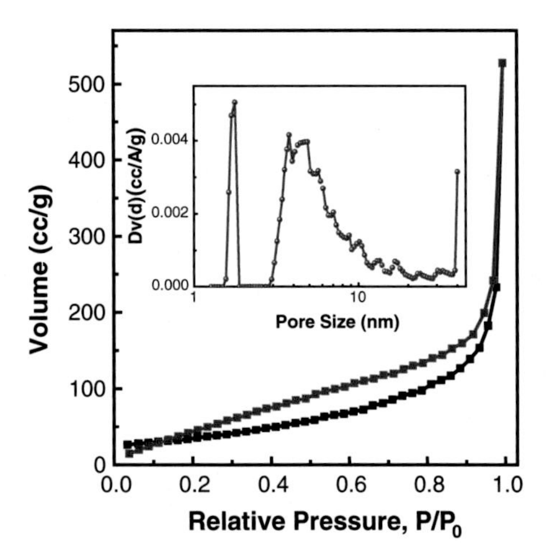

Fig. 5.4 N_2 physic sorption isotherm (*Inset* shows the DFT pore size distribution) of MGF. Reproduced from Ref. [38] by permission of The Royal Society of Chemistry

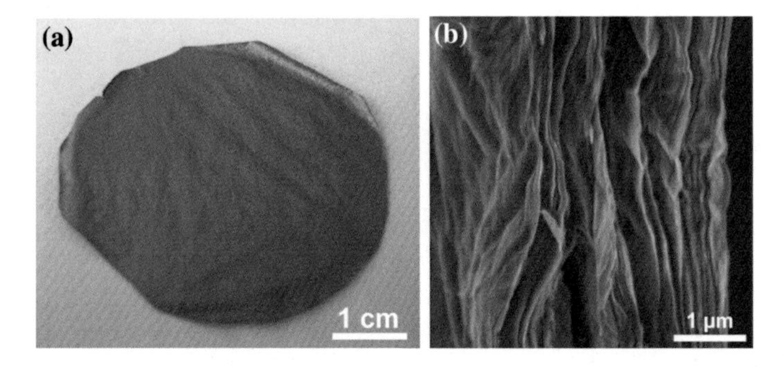

Fig. 5.5 **a** Digital photo of a compact graphene film (CGF) fabricated by a routine filtration method. **b** SEM image of the cross-section of a CGF. Reproduced from Ref. [38] by permission of The Royal Society of Chemistry

basic building blocks of MGF while the bubble wall was constructed by few layer graphene nanosheets.

Thermogravimetric (TG) analysis was employed to investigate the structural evolution of GO/PMMA composite film during calcination (Fig. 5.3a). The oxygen-containing functional groups of GO were mainly removed at 230 °C, indicating a fast thermal reduction toward thermally derived graphene. The decompose of PMMA spheres started at 250 °C. After 400 °C, PMMA were totally removed. In the case of the GO/PMMA composite film, two stages of weight loss at 230 and 380 °C were attributed to the thermal reduction of GO and pyrolysis of PMMA, respectively. The thermally derived graphene was formed with hollow macroporous

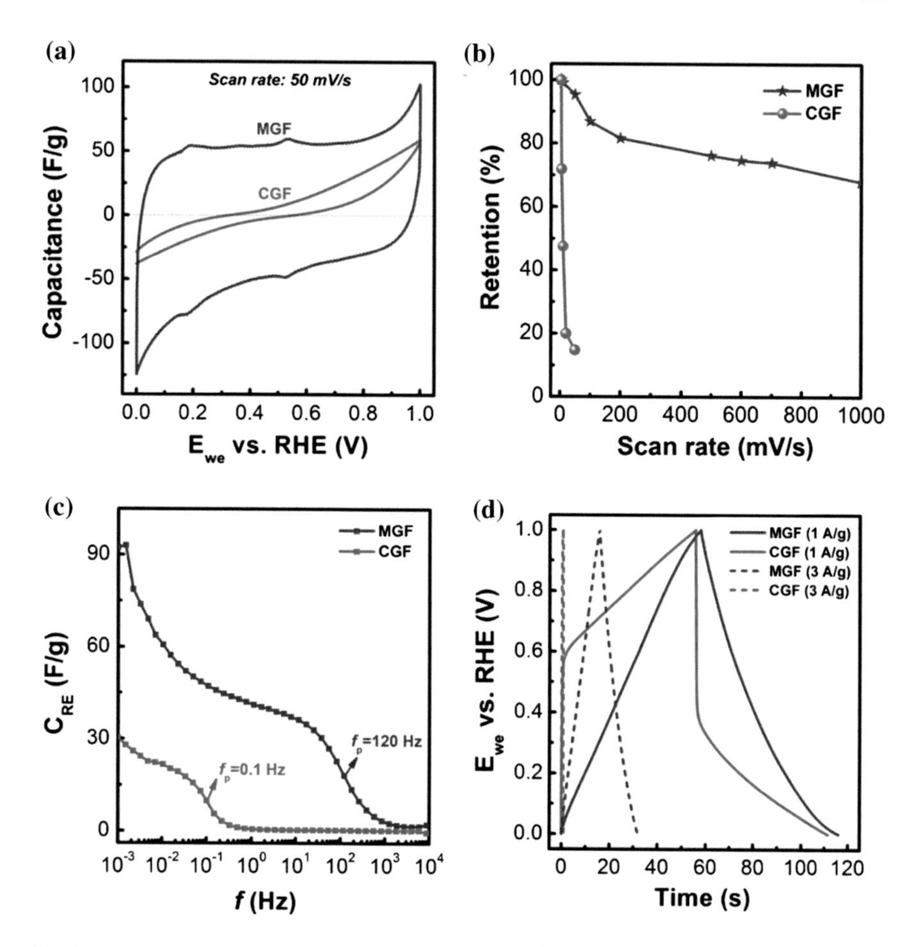

Fig. 5.6 **a** CVs of MGF and CGF at scan rate of 50 mV s^{-1}. **b** The rate retention of MGF and CGF. **c** Real part of complex capacitance as a function of AC frequency by EIS. **d** GC curves at current density of 1 and 3 A g^{-1} of MGF and CGF. Reproduced from Ref. [38] by permission of The Royal Society of Chemistry

bubbles serving as the interlayer spacer. A residual weight of 7.6 wt% was observed, which comes from thermally derived graphene, since the sacrificed PMMA spheres have been satisfactorily replicated after high temperature calcination.

Both Hg penetration and N_2 physic-sorption were employed to characterize the pore structure of the as-prepared MGF. The Hg penetration result represented a very high porosity (91.05 %) with an ultra-low bulk density of 0.0849 g mL^{-1}, indicating that the film was very light. The specific surface area deduced from N_2 sorption (BET model, micro-, and mesopores) was determined to be 128.2 m^2 g^{-1}. The macroporous size distribution from Hg penetration analysis was shown in Fig. 5.3b, with the

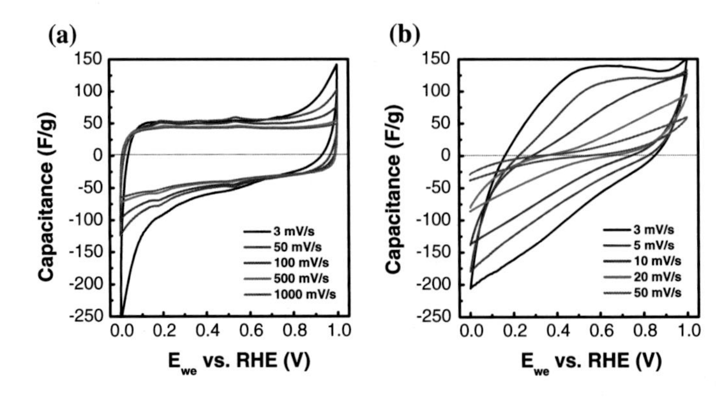

Fig. 5.7 **a** CV evolution of MGF at different scan rates from 3 to 1000 mV s^{-1}. **b** CV evolution of CGF at different scan rates from 3 to 50 mV s^{-1}. Reproduced from Ref. [38] by permission of The Royal Society of Chemistry

average pore diameter of 107.3 nm. The mesopores among the graphene nanosheets can be observed on the DFT pore size distribution from N_2 sorption results (Fig. 5.4).

The as-obtained graphene film provides mediated 3D macropores for ion-buffering reservoirs and low-resistant channels for ion diffusion, which are promising characteristics for a high rate supercapacitor electrode. Cyclic voltammetry (CV), electrochemical impedance spectroscopy (EIS) and Galvanostatic charge-discharge (GC) was employed to evaluated the capacitive performance of MGF (Fig. 5.6), with CGF as the control sample (Fig. 5.5) [14]. As shown in Fig. 5.6a, MGF exhibits a typical electrochemical double layer capacitive (EDLC) behavior. Interestingly, the profiles of CV keep rectangular with the increase of scan rate (from 3 to 1000 mV s^{-1}) (Fig. 5.7a). The calculated capacitance thus revealed high rate retention (67.9 %) at 1000 mV s^{-1} (Fig. 5.6b). Conversely, seldom capacitive characteristic was observed for CGF, which only exhibits very narrow CV curve at 50 mV s^{-1} (Fig. 5.7b). Based on a series of RC model, the capacitance (CRE) plots from a frequency response analysis (FRA) [39] are presented in Fig. 5.6c. A monotonic increase from 0 to maximum value was observed with the decrease of AC frequency, which indicates the restriction of ion transport at high frequency. The utilizable total capacitance of MGF and CGF electrodes were determined to be 92.7 and 30.6 F g^{-1}, respectively. In addition, the peak frequency (f_P) of MGF (1500 Hz) is much higher than that of CGF (0.5 Hz), which shows a significantly better rate capability for MGF. The frequency response results reveal the significant influence of the porous structures on the rate of ion transport [40].

Clearly, the opened macroporous structure of MGF provide a convenient ion diffusion channel, which is unavailable in CGF with an over packed morphology. Supercapacitor performance was also characterized by GC curves (Fig. 5.6b) and yields a similar specific capacitance of around 58 F g^{-1} for MGF with a constant current of 1.0 A g^{-1}. As the current density increases to 3.0 A g^{-1}, MGF still owns a specific capacitance of 49.2 F g^{-1}. However, the performance of CGF deteriorated very rapidly with little capacitance left. Furthermore, a large IR_{drop} (0.6 V) can be observed for CGF at 1.0 A g^{-1} than that of MGF which has a symmetrical pattern of GC. This also indicates that the huge equivalent series resistance of CGF is due to the compact structure, which is impermeable to the electrolyte. Thus, the better high rate performance of three-dimensional bubble graphene film is attributed to its novel pore structure and excellent electric conductivity.

5.4 Conclusions

A 3D MGF, with hollow graphene bubble serving as the replicated building block, was obtained by a facile hard template-directed ordered assembly approach. The graphene oxide precursor and template PMMA spheres were orderly assembled by vacuum filtration, followed by calcination to reduce graphene oxide into graphene, as well as to remove the template via pyrolysis. The obtained free-standing film showed quite high rate capability (1.0 V s^{-1}) working as 'binder-free' supercapacitor electrode. This work provides a simple and green synthetic strategy to obtain free-standing macroporous graphene assemblies with tunable microstructure, which shows promising prospects towards applications in biology, energy conversion, and catalysis, etc.

References

1. Zhai YP, Dou YQ, Zhao DY, Fulvio PF, Mayes RT, Dai S. Carbon materials for chemical capacitive energy storage. Adv Mater. 2011;23(42):4828–50.
2. Liu R, Duay J, Lee SB. Heterogeneous nanostructured electrode materials for electrochemical energy storage. Chem Commun. 2011;47(5):1384–404.
3. Su DS, Schlogl R. Nanostructured carbon and carbon nanocomposites for electrochemical energy storage applications. ChemSusChem. 2010;3(2):136–68.
4. Xu Z, Gao C. Graphene chiral liquid crystals and macroscopic assembled fibres. Nat Commun. 2011;2:1–9.
5. Chen C, Yang Q-H, Yang Y, Lv W, Wen Y, Hou P-X, et al. Self-assembled free-standing graphite oxide membrane. Adv Mater. 2009;21(29):3007–11.
6. Dikin DA, Stankovich S, Zimney EJ, Piner RD, Dommett GHB, Evmenenko G, et al. Preparation and characterization of graphene oxide paper. Nature. 2007;448(7152):457–60.
7. Ramanathan T, Abdala AA, Stankovich S, Dikin DA, Herrera-Alonso M, Piner RD, et al. Functionalized graphene sheets for polymer nanocomposites. Nat Nanotechnol. 2008;3(6): 327–31.
8. Stankovich S, Dikin DA, Dommett GHB, Kohlhaas KM, Zimney EJ, Stach EA, et al. Graphene-based composite materials. Nature. 2006;442(7100):282–6.
9. Xiao J, Mei DH, Li XL, Xu W, Wang DY, Graff GL, et al. Hierarchically porous graphene as a lithium-air battery electrode. Nano Lett. 2011;11(11):5071–8.
10. Worsley MA, Pauzauskie PJ, Olson TY, Biener J, Satcher JH Jr, Baumann TF. Synthesis of graphene aerogel with high electrical conductivity. J Am Chem Soc. 2010;132(40):14067–9.

11. Lee SH, Kim HW, Hwang JO, Lee WJ, Kwon J, Bielawski CW, et al. Three-dimensional self-assembly of graphene oxide platelets into mechanically flexible macroporous carbon films. Angew Chem Int Ed. 2010;49(52):10084–8.
12. Chen Z, Ren W, Gao L, Liu B, Pei S, Cheng H-M. Three-dimensional flexible and conductive interconnected graphene networks grown by chemical vapour deposition. Nat Mater. 2011;10(6):424–8.
13. An SJ, Zhu YW, Lee SH, Stoller MD, Emilsson T, Park S, et al. Thin film fabrication and simultaneous anodic reduction of deposited graphene oxide platelets by electrophoretic deposition. J Phys Chem Lett. 2010;1(8):1259–63.
14. Chen C-M, Huang J-Q, Zhang Q, Gong W-Z, Yang Q-H, Wang M-Z, et al. Annealing a graphene oxide film to produce a free standing high conductive graphene film. Carbon. 2012;50(2):659–67.
15. Wei W, Lu W, Yang QH. High-concentration graphene aqueous suspension and a membrane self-assembled at the liquid-air interface. New Carbon Mater. 2011;26(1):36–40.
16. Lv W, Xia ZX, Wu SD, Tao Y, Jin FM, Li BH, et al. Conductive graphene-based macroscopic membrane self-assembled at a liquid-air interface. J Mater Chem. 2011;21(10):3359–64.
17. Tang ZH, Zhuang J, Wang X. Exfoliation of graphene from graphite and their self-assembly at the oil-water interface. Langmuir. 2010;26(11):9045–9.
18. Fan ZJ, Yan J, Wei T, Ning GQ, Zhi LJ, Liu JC, et al. Nanographene-constructed carbon nanofibers grown on graphene sheets by chemical vapor deposition: high-performance anode materials for lithium ion batteries. ACS Nano. 2011;5(4):2787–94.
19. Fan ZJ, Yan J, Zhi LJ, Zhang Q, Wei T, Feng J, et al. A three-dimensional carbon nanotube/graphene sandwich and its application as electrode in supercapacitors. Adv Mater. 2010;22(33): 3723–8.
20. Tung VC, Chen L-M, Allen MJ, Wassei JK, Nelson K, Kaner RB, et al. Low-temperature solution processing of graphene-carbon nanotube hybrid materials for high-performance transparent conductors. Nano Lett. 2009;9(5):1949-55F.
21. Wu Z-S, Ren W, Wen L, Gao L, Zhao J, Chen Z, et al. Graphene anchored with Co_3O_4 nanoparticles as anode of lithium ion batteries with enhanced reversible capacity and cyclic performance. ACS Nano. 2010;4(6):3187–94.
22. Si Y, Samulski ET. Exfoliated graphene separated by platinum nanoparticles. Chem Mater. 2008;20(21):6792–7.
23. Wang G, Wang B, Wang X, Park J, Dou S, Ahn H, et al. Sn/graphene nanocomposite with 3D architecture for enhanced reversible lithium storage in lithium ion batteries. J Mater Chem. 2009;19(44):8378–84.
24. Mai YJ, Wang XL, Xiang JY, Qiao YQ, Zhang D, Gu CD, et al. CuO/graphene composite as anode materials for lithium-ion batteries. Electrochim Acta. 2011;56(5):2306–11.
25. Yan J, Wei T, Shao B, Ma F, Fan Z, Zhang M, et al. Electrochemical properties of graphene nanosheet/carbon black composites as electrodes for supercapacitors. Carbon. 2010;48(6): 1731–7.
26. Yang X, Zhu J, Qiu L, Li D. Bioinspired effective prevention of restacking in multilayered graphene films: towards the next generation of high-performance supercapacitors. Adv Mater. 2011;23(25):2833–8.
27. Eswaraiah V, Sankaranarayanan V, Ramaprabhu S. Functionalized graphene-PVDF foam composites for EMI shielding. Macromol Mater Eng. 2011;296(10):894–8.
28. Worsley MA, Olson TY, Lee JRI, Willey TM, Nielsen MH, Roberts SK, et al. High surface area, sp^2-cross-linked three-dimensional graphene monoliths. J Phys Chem Lett. 2011;2(8): 921–5.
29. Lin Y, Ehlert GJ, Bukowsky C, Sodano HA. Superhydrophobic functionalized graphene aerogels. ACS Appl Mater Interf. 2011;3(7):2200–3.
30. Chen WF, Li SR, Chen CH, Yan LF. Self-assembly and embedding of nanoparticles by in situ reduced graphene for preparation of a 3D graphene/nanoparticle aerogel. Adv Mater. 2011;23(47):5679–83.

31. Zhang X, Sui Z, Xu B, Yue S, Luo Y, Zhan W, et al. Mechanically strong and highly conductive graphene aerogel and its use as electrodes for electrochemical power sources. J Mater Chem. 2011;21(18):6494–7.
32. Vickery JL, Patil AJ, Mann S. Fabrication of graphene-polymer nanocomposites with higher-order three-dimensional architectures. Adv Mater. 2009;21(21):2180–4.
33. Liu F, Seo TS. A controllable self-assembly method for large-scale synthesis of graphene sponges and free-standing graphene films. Adv Funct Mater. 2010;20(12):1930–6.
34. Shao JJ, Wu SD, Zhang SB, Lv W, Su FY, Yang QH. Graphene oxide hydrogel at solid/liquid interface. Chem Commun. 2011;47(20):5771–3.
35. Guo X, Zhang F, Evans DG, Duan X. Layered double hydroxide films: synthesis, properties and applications. Chem Commun. 2010;46(29):5197–210.
36. Yang XW, He YS, Liao XZ, Ma ZF. Improved graphene film by reducing restacking for lithium ion battery applications. Acta Phys-Chim Sin. 2011;27(11):2583–6.
37. Fan D, Liu Y, He J, Zhou Y, Yang Y. Porous graphene-based materials by thermolytic cracking. J Mater Chem. 2012;22(4):1396–402.
38. Chen CM, Zhang Q, Huang CH, Zhao XC, Zhang BS, Kong QQ, et al. Macroporous 'bubble' graphene film *via* template-directed ordered-assembly for high rate supercapacitor. Chem Commun. 2012;48:7149–51.
39. Jang JH, Kato A, Machida K, Naoi K. Supercapacitor performance of hydrous ruthenium oxide electrodes prepared by electrophoretic deposition. J Electrochem Soc. 2006;153(2):A321–8.
40. Zhu Y, Murali S, Stoller MD, Ganesh KJ, Cai W, Ferreira PJ, et al. Carbon-based supercapacitors produced by activation of graphene. Science. 2011;332(6037):1537–41.

Chapter 6
SnO_2@Graphene Composite Electrodes for the Application in Electrochemical Energy Storage

6.1 Introduction

Advanced energy storage and conversion is of great importance for the challenge of global warming and the finite nature of fossil fuels. Electrochemical energy storage devices, such as supercapacitors and batteries, are playing a core role in balancing the energy generated by engines, solar, and wind power. Supercapacitors, or electrochemical capacitors, are energy storage devices that store charges electrostatically through the reversible adsorption/desorption of ions in the electrolyte onto active materials [1, 2], while Li-ion batteries, which consist of two electrodes that are capable of reversibly hosting Li in ionic form [3–5], are widely used for consumer electronics, power management, and hybrid electric vehicles. However, the energy density or power density of the two systems still needs to be improved for demanding applications. Exploring advanced electrode materials is one of the most straightforward approaches to improve the efficiency of electrochemical energy storage systems.

The element carbon is a very flexible choice for building electrochemical energy storage devices [6, 7]. Graphite, glassy carbon, carbon black, hard carbon, mesocarbon microbead, activated carbon, and mesoporous carbon are widely utilized in energy storage systems [8, 9]. Carbon nanotubes have been applied as an electric conductive additive for Li-ion batteries [10]. Graphene, the two-dimensional carbon crystal lattice with excellent electronic conductivity, optical transparency, mechanical strength, inherent flexibility, and huge theoretical surface area, has been considered a novel and portable nanocarbon component for electrodes of energy storage devices [11–17]. However, the irreversible aggregation of graphene sheets always creates more void units and masks active sites for pseudocapacitor or Li-intercalation. Thus, metal oxides (e.g., RuO_2 [18], MnO_2 [1, 19, 20], Co_3O_4 [21–23], NiO [24], SnO_2 [25–35], Fe_3O_4 [36]), conductive polymers (e.g., PANI [37, 38], PPy [39]) or nanocarbon (e.g., carbon nanotube [40]) are introduced as secondary phases that serve as 'spacers' to avoid the

C.-M. Chen, *Surface Chemistry and Macroscopic Assembly of Graphene for Application in Energy Storage*, Springer Theses,
DOI 10.1007/978-3-662-48676-4_6

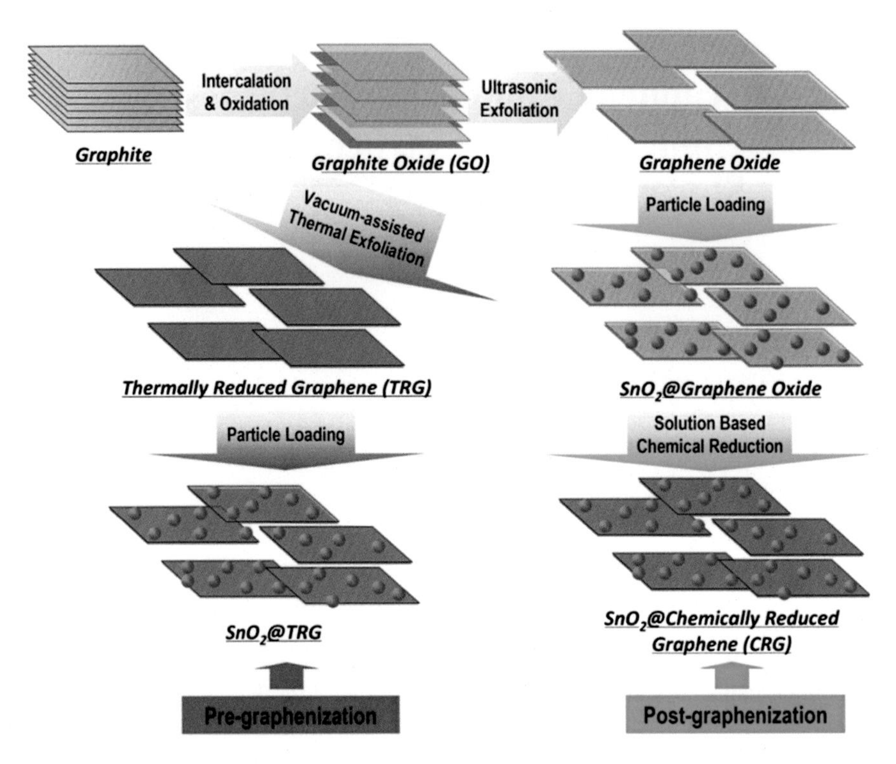

Fig. 6.1 Illustration of the pre- and post-graphenization route for metal oxide/graphene hybrid fabrication. Reproduced from Ref. [41] by permission of The Royal Society of Chemistry

restacking of graphene, provide pseudo-active centers for supercapacitors, or host storage materials for Li-ion batteries. The synergistic effects between graphene and the secondary phase such as space confinement, electronic modification, and fast charge transfer will also enhance the electrochemical performance of the electrode materials. The applications of graphene-based electrodes for supercapacitor and Li-ion battery use are also highlighted by recent reviews [12–14].

To develop graphene-based hybrid electrodes with superior performance in electrochemical energy storage, the selection of appropriate raw materials and the optimization of synthesis strategy are important. Among various approaches for graphene production, the chemically derived graphene (CDG), which uses graphite, graphite oxide (GO) or other graphite derivatives as starting materials, can be produced in large scale, and provides further processability and abundant functions for industrial applications as well [11–14]. Up to now, the reduction of GO to CDGs is the most widely applied technique for the large-scale preparation of graphene. During the preparation of graphene-based hybrids, as shown in Fig. 6.1, the current reported methods can be simply classified into two general strategies according to the processes of CDG synthesis: (1) pre-graphenization strategy: CDG is synthesized [such as thermally reduced graphene (TRG)] before the second component is introduced; (2) post-graphenization strategy: a composite composed of a CDG precursor (usually graphene oxide) and the second component is preprepared, followed by converting the precursor into chemical reduced graphene (CRG). Both of them have been widely used for graphene-based hybrid electrode fabrication, but it is still an open question as to which one is better.

To explore an advanced technique as well as reveal the chemical and material science for fabrication of graphene-based electrode materials, SnO_2 is selected as the secondary component to the graphene supports by pre- or post-graphenization. Both pre-graphenization and post-graphenization process are carried out for SnO_2 @graphene hybrid fabrication, as shown in Fig. 6.1, and the obtained SnO_2@TRG and SnO_2@CRG electrodes are carefully characterized. Both the graphenes and their hybrids are evaluated for supercapacitors and Li-ion battery electrodes, so as to provide insightful materials chemistry toward developing advanced graphene-based electrodes.

6.2 Experimental

6.2.1 Pre-graphenization: SnO_2@TRG Hybirds

The thermally reduced graphene was obtained by thermal expansion of GO powder under high vacuum. GO was prepared by a modified Hummers method. The as-prepared GO was put into a quartz tube that was sealed at one end and stoppered at the other end, through which the reactor was connected to the high vacuum pump. The tube was heated at a rate of 30 °C min^{-1} under high vacuum (<3.0 Pa). At about 200 °C, an abrupt expansion was observed. To remove the abundant functional groups, the expanded GO was kept at 250 °C for 20 min and a high vacuum was maintained (<5.0 Pa) during heat treatment. The as-prepared graphene sample was denoted as TRG. The SnO_2@TRG hybrids were prepared by a facile excessive impregnation of the above TRG. In a typical process, 500 mg of TRG was mixed with 500 mL of 0.1 M $SnCl_2$ aqueous solution (with 0.1 M HCl as pH adjuster) in ice bath (~0 °C). To realize moderate anchoring of Sn^{2+} on the active sites of graphene, the black mixture was kept in the ice bath with intensive stirring for 10 min. The product was isolated by vacuum filtration and rinsed copiously with water (5 × 100 mL) and ethanol (5 × 100 mL). Finally, the sample was air dried at 110 °C for 24 h to obtain SnO_2@TRG hybrids.

6.2.2 Post-graphenization: SnO_2@CRG Hybrids

The CRG was prepared by a chemical reduction approach. In a typical procedure, GO (500 mg) was dispersed in 500 mL water followed by sonication (200 W) for

30 min to yield a homogeneous brown hydrosol of graphene oxide. The above hydrosol was mixed with 50 mL hydrazine monohydrate (NH_2–NH_2·H_2O, 100 %) in a 1000 mL round-bottom flask, and heated in an oil bath at 100 °C under a water-cooled condenser for 24 h, during which the reduced GO gradually precipitated out as black solids. The product was isolated by vacuum filtration and washed thoroughly with water and ethanol to remove excessive metal salts. Finally, the sample was air dried in a watch glass at 110 °C for 24 h, obtaining chemically reduced graphene.

The SnO_2@CRG hybrids were prepared by pre-impregnation of graphene oxide followed by a similar chemical reduction approach. In a typical experimental, 500 mL of graphene oxide hydrosol (1.0 mg mL^{-1}) was premixed with 250 mL of $SnCl_2$ aqueous solution. The mixture was stirred in ice bath for 10 min, and then moved to oil bath over which 50 mL of 100 % NH_2–NH_2·H_2O was added gradually. The solution was heated and kept at 100 °C with intensive stirring under a water-cooled condenser for 24 h. The obtained SnO_2@CRG hybrids are separated, washed, and dried using the same method as mentioned above for CRG.

6.2.3 Sample Characterization

The morphologies were characterized using Hitachi S4800 scanning electron microscope (SEM) operated at 2.0 kV and Philips CM200 LaB 6 transmission electron microscope (TEM) operated at 200.0 kV. The HRTEM images of the samples were collected on FEI Cs-corrected Titan 80–300 microscope operated at 80.0 kV. Energy dispersive X-ray spectroscopy (EDS) analysis was performed using Titan 80–300 apparatus with the analytical software INCA. X-ray photoelectron spectroscopy (XPS) was performed on the Thermo VG ESCA-LAB250 surface analysis system. Brunauer–Emmett–Teller (BET) specific surface areas and Barret-Joyner-Halenda (BJH) pore size distributions of the SnO_2@graphene hybrids were determined by N_2 physisorption at 77 K using Micromeritics 2375 BET apparatus.

6.2.4 Li-Ion Battery Performance Measurements

The performances of graphene-based hybrids as anode materials for lithium ion batteries were tested with CR2025 coin cells. A mixture of SnO_2@CRG nanocomposites or SnO_2@TRG, carbon black, and polyvinylidene fluoride at a weight ratio of 80:10:10 were pasted on pure Cu foil (99.6 %, Goodfellow) to make the working electrode. A microporous polyethylene sheet (Celgard 2400) was used as separator. The electrolyte was 1.0 M $LiPF_6$ dissolved in a mixed solution of ethylene carbonate–dimethyl carbonate–ethylene methyl carbonate (1:1:1, by weight) obtained from Ube Industries Ltd. Pure lithium foil (Aldrich) was used as counter electrode. The cells were assembled in an Ar-filled glove box. The discharge and charge measurements were carried out at different current densities in the voltage range of 0–3.0 V on a Neware battery test system. The specific capacity of the SnO_2@graphene nanocomposites was calculated based on the mass of the anode materials (SnO_2 and graphene). Cyclic voltammogram measurements were performed on a Solartron 1470E electrochemical workstation at a scan rate of 0.1 mV s^{-1}.

6.2.5 Supercapacitor Performance Measurements

The electrochemical properties of graphene-SnO_2 hybrids were measured in aqueous system (electrolyte: 6.0 M KOH). A three-electrode system was employed in the measurement, whereby Ni foam coated with electrode materials served as the working electrode, a platinum foil electrode as counter electrode and a saturated hydrogen electrode (SHE) served as reference electrode. In order to prepare a working electrode, a mixture of our active material, carbon black, and poly(tetrafluoroethylene) with a weight ratio of 80:5:15 was ground together to form a homogeneous slurry. The slurry was squeezed into a film and then punched into pellets. The punched pellets with a piece of nickel foam on each side were pressed under 2.5 MPa and dried overnight at 110 °C. Each electrode was quantified to contain roughly 5.0 mg active materials. The electrodes were impregnated with electrolyte under vacuum for 1.0 h prior to the electrochemical evaluation. Cyclic voltammogram (CV) curves (scan rates varying from 3 to 500 mV s^{-1}) and electrochemical impedance spectroscopy (EIS) profiles were measured with VSP BioLogic electrochemistry workstation. The electrochemical capacitances were both obtained from CV curves. The Nyquist plot was fitted by EC-Lab software.

6.3 Results and Discussion

6.3.1 Pre-graphenization: Structure of TRG and SnO_2@TRG Hybrids

In the pre-graphenization process, the GO is thermally exfoliated and reduced into fluffy black powder [42]. As shown in Fig. 6.2a, the as-prepared TRGs have a hierarchically honeycomb-like morphology: a continuous and interconnected 3D macroscopic architecture, with a Brunauer–Emmett–Teller (BET) surface area of 293 m^2 g^{-1}, is constructed by crumpled graphene sheets which are loosely stacked or folded with each other. The diameter of the macropores ranges from 100 to 300 nm (Fig. 6.2a, b). The TRG has a coarse edge, which can be attributed to the fast decomposition of oxygen-containing functional groups (such as carboxyl, carbonyl groups) during thermal exfoliation. After impregnation and calcination process, the pre-graphenized SnO_2@TRG hybrids still remain a porous

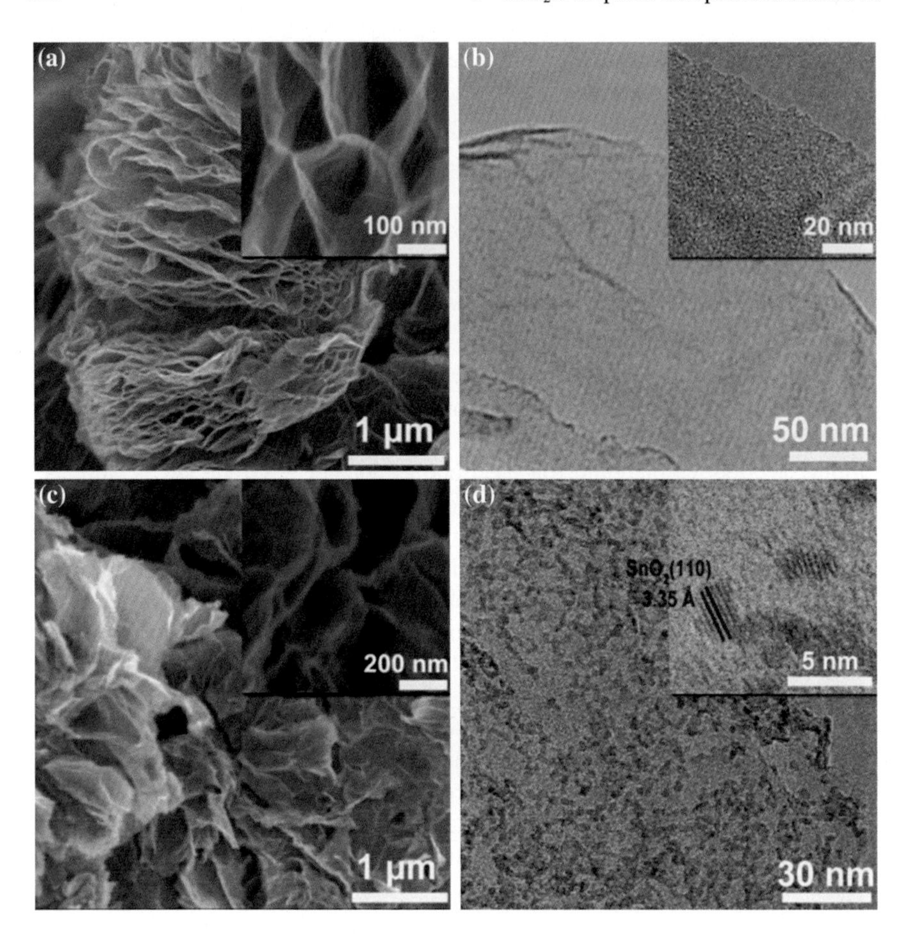

Fig. 6.2 **a** SEM image of TRG; **b** TEM image of TRG; **c** SEM image of SnO_2@TRG; **d** TEM image of SnO_2@TRG. Reproduced from Ref. [41] by permission of The Royal Society of Chemistry

morphology with twisted and loosely packed graphene sheets. However, the regularity of the ordered honeycomb structure is lost gradually. As the basic building blocks, the graphene sheets are larger than 1 μm in diameter, between which pores ranging from 20 to 100 nm are clearly identified (Fig. 6.2c). The microtexture of SnO_2@TRG was further examined by TEM, most SnO_2 nanoparticles with sizes ranging between 3 and 5 nm with a density of ca. 4×10^{12} cm^{-2} are uniformly distributed on the graphene sheets. Some of the SnO_2 particles are agglomerated into chain-like structures. As shown in the inserted figure of the high resolution TEM images (Fig. 6.2d), the lattice with a d_{110} space of 3.35 Å can be clearly demonstrated, indicating good crystallization of SnO_2 [26].

The fine scan X-ray diffraction (XRD) (collected at a very slow scan rate of 0.1° min^{-1}) was employed to determine the crystallite structure of pre-graphenized samples. As shown in Fig. 6.3a, after thermal graphenization, TRG exhibits a tiny sharp peak (002) with a broad shoulder at 26.6° and a very weak peak (101) at 43.6°. The sharp peak is attributed to the thin-layer graphitic microcrystalline stacking by graphene layers due to inadequate exfoliation of graphite oxide during thermal expansion, while the shoulder is ascribed to the local positive fluctuation of interlayer spacing of graphene layers due to the rotation, translation, curvature, and fluctuation of atomic positions along the normal of graphene layers [43]. After SnO_2 introduction, despite the carbon-related peaks, several very weak peaks around 33.2°, 51.7°, 64.0° emerge, which are assigned to the (101), (211) and (310) lines of the newly formed α-SnO_2 phase, respectively, while the increased intensity of the peak and shoulder at 26.6° is ascribed to the restacking of graphene during impregnation and drying, and the amalgamation of the (002) line of graphite with the nearby (110) line of SnO_2. The crystalline size along the (110) lattice of SnO_2 is estimated to be 3.8 nm according to the Scherrer equation, which is consistent with the TEM observation.

The N_2 adsorption–desorption isotherms are carried out to evaluate the change of pore structure for TRG before and after SnO_2 impregnation. As shown in Fig. 6.2c, both adsorption isotherms exhibit the typical Type III isotherm with H3 hysteresis loop according to IUPAC classification, indicating materials characteristic of macropores (pore size >50 nm) and comprised of aggregates (loose assemblages) of plate-like particles forming slit-like mesopores (Fig. 6.3b). The specific BET surface area (S_{BET}), *t*-plot micropore surface area (pore size <2 nm) (S_{micro}), and total pore volume (V_P) of TRG and SnO_2@TRG are summarized in Table 6.1. After impregnation, the S_{BET} is slightly increased from 293.0 to 324.6 $m^2\ g^{-1}$, while the V_P is decreased simultaneously from 1.62 to 1.26 $cm^3\ g^{-1}$, with a corresponding increase of S_{micro} from 6.9 to 12.8 $m^2\ g^{-1}$. The BJH adsorption pore size distribution (Fig. 6.3c) indicates the key role of exterior meso- and macropores with relatively larger diameters (pore size >2 nm) in contributing the S_{BET} of TRG and SnO_2@TRG, however, the average diameter of pore decreased after SnO_2 impregnation. The evolution of porous structure is attribute to the collapse of exterior macropores for the capillary effect during the air drying process (Fig. 6.3c).

Furthermore, the XPS results show that the atomic percentages of Sn, C, and O element are 1.71, 85.84, and 12.44 at.%, respectively (Table 6.1). From the fine scan of SnO_2@TRG (Fig. 6.3d), the Sn components present the typical $3d_{5/2}$ (485.6 eV) and $3d_{3/2}$ (494.0 eV) level with a gap of 8.4 eV and area ratio of 1.5, which further confirms the state of SnO_2 on TRG [25, 30].

6.3.2 Post-graphenization: Structure of SnO_2@CRG Hybrids

The other route for SnO_2@graphene hybrids is post-graphenization. The Sn^{4+} species are loaded onto the surface of GO ahead of time. Then the GO is reduced into graphene by solution-based chemical reduction. As shown in Fig. 6.4a, the CRG

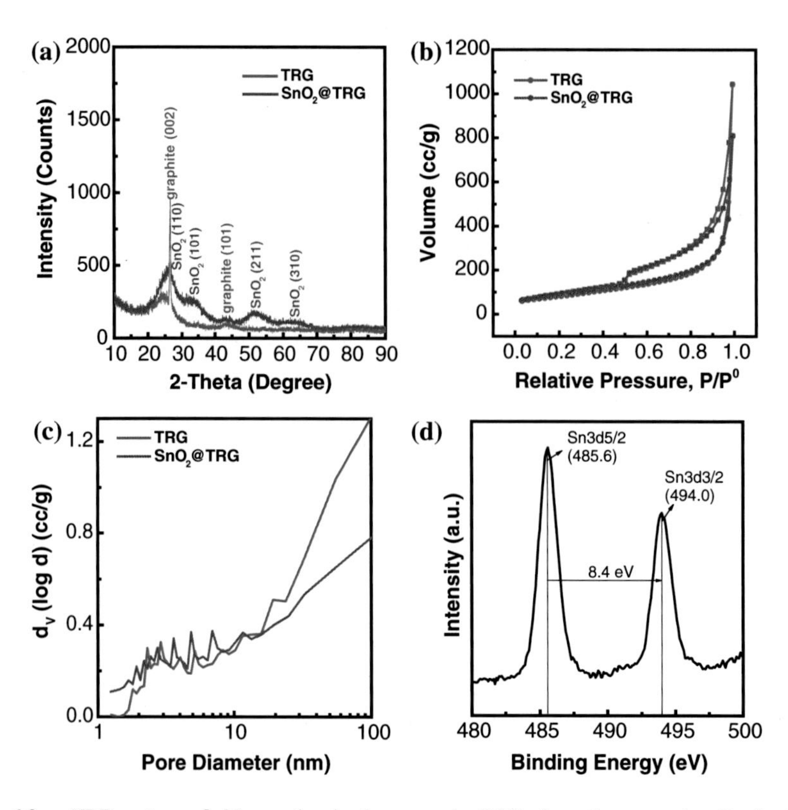

Fig. 6.3 **a** XRD patterns, **b** N_2 sorption isotherms and **c** BJH adsorption pore size distributions of TRG and SnO_2@TRG; **d** XPS Sn3d fine scan spectrum of SnO_2@TRG. Reproduced from Ref. [41] by permission of The Royal Society of Chemistry

Table 6.1 Quantification Results of BET, XRF and XPS

Sample	$S^{a}_{BET}(m^2/g)$	$V^{a}_{P}(m^3/g)$	$S^{a}_{micro}(m^2/g)$	Bulk. Sn[b] (%)	Surf. Sn[c] (%)	Surf. C[c] (%)	Surf. O[c] (%)
TRG	293.0	1.62	6.9	0	0.00	89.70	10.30
SnO_2@TRG	324.6	1.26	12.8	3.86	1.71	85.84	12.44
CRG	666.3	0.60	69.9	0	0.00	88.43	11.57
SnO_2@CRG	818.4	1.09	NA	2.00	0.74	88.06	11.21

Reproduced from Ref. [41] by permission of The Royal Society of Chemistry
[a]Calculated by N_2 physisorption
[b]Obtained from XRF
[c]Quantified by XPS

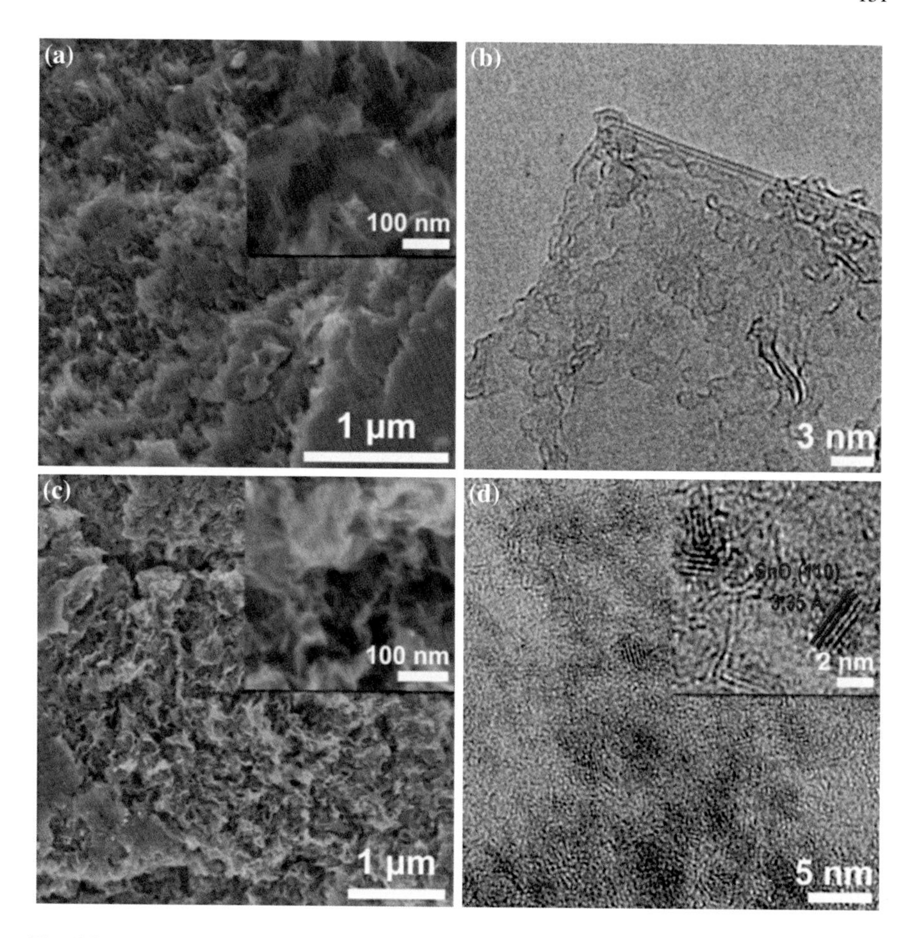

Fig. 6.4 **a** SEM and **b** TEM images of CRG; **c** SEM and **d** TEM images of SnO_2@CRG. Reproduced from Ref. [41] by permission of The Royal Society of Chemistry

exhibits closely packed morphology, in which the graphene sheets are randomly crumpled (Fig. 6.4b). After SnO_2 loading, the post-graphenized SnO_2@CRG hybrid still holds a highly twisted structure (Fig. 6.4c), in which only micropores and mesopores are observed. SnO_2 nanoparticles with a diameter of 3.5–5.5 nm and a density of ca. 3×10^{12} cm^{-2} are uniformly distributed on graphene sheets. The lattice of SnO_2 (d_{110} ~ 3.35 Å) nanoparticles can be clearly observed in the HRTEM images (Fig. 6.4d), which indicates high crystallization degree.

Comparing with the pre-graphenized TRG, CRG exhibits broader and stronger diffraction peaks of the graphite (002) and (101) lattice in the XRD pattern, indicating an over re-stacking and entanglement of graphene sheets within the macro-assembly. However, after introduction of Sn components as low as 2 wt%, the graphite associated peaks are significantly minimized while very weak SnO_2 peaks

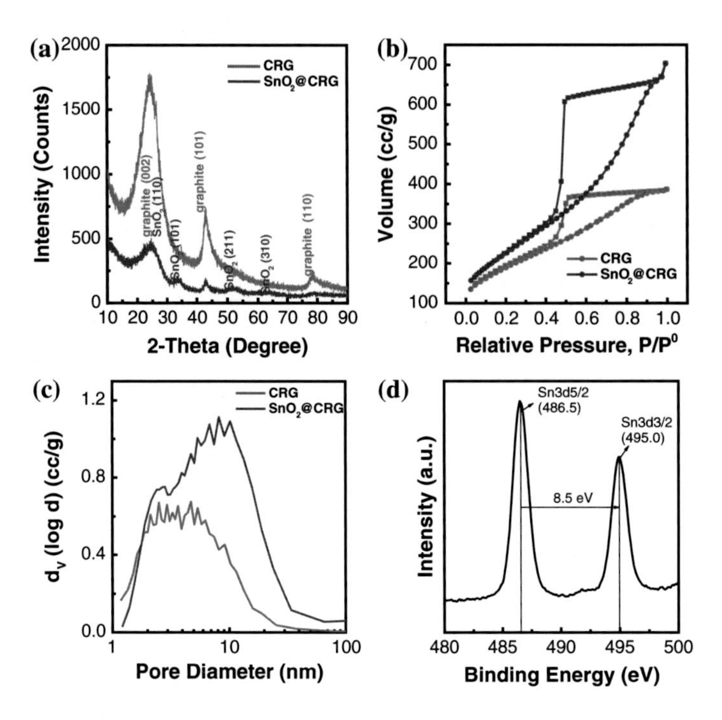

Fig. 6.5 **a** XRD patterns, **b** N_2 sorption isotherms and **c** BJH adsorption pore size distributions of CRG and SnO_2@CRG; **d** XPS Sn3d fine scan spectrum of SnO_2@CRG. Reproduced from Ref. [41] by permission of The Royal Society of Chemistry

appear, due to the formation of SnO_2 nanocrystals acting as the isolation spacers between graphene sheets. Thus, the BET specific surface area is increased from 666 $m^2 g^{-1}$ for CRG to 818.4 $m^2 g^{-1}$ for SnO_2@CRG. It might be attributed to more closed pores turn to opened ones which are accessible to the N_2 molecules (Table 6.1) [43].

As shown in Fig. 6.5b, the N_2 sorption isotherms of both CRG and SnO_2@CRG present a typical Type IV isotherm with apparent H2 hysteresis loop, which indicates the presence of ink-bottle pores in SnO_2@CRG. It is interesting that the curve of N_2 sorption isotherm of CRG is apparently different from that of TRG while no macropores are found in CRG. Furthermore, the S_{BET} of CRG (666.3 $m^2 g^{-1}$) is much higher than that of TRG (293.0 $m^2 g^{-1}$). These phenomena are mainly ascribed to more mesopores generated by over restacking and entanglement of graphene sheets in CRG, and to poorly exfoliation of graphite oxide for TRG. After introduction of SnO_2, the pore volume is increased from 0.60 to 1.09 $cm^3 g^{-1}$ and the vanishing micropore surface area in SnO_2@CRG is observed. The BJH adsorption pore size distribution of CRG and SnO_2@CRG exhibits the parabolic profile with peaks at 4 and 10 nm, respectively, which is different from pre-graphenized SnO_2@TRG samples.

As calculated from XPS, the surface contents of Sn, C, and O on SnO_2@CRG are determined to be 0.74, 88.06, and 11.21 at.%, respectively (Table 6.1). The Sn3d fine scan spectrum exhibits the similar $3d_{5/2}$ and $3d_{3/2}$ line to SnO_2@TRG at 486.5, 495.0 eV with a gap of 8.5 eV, indicating the grafting of SnO_2 on CRG. However, the content of Sn in SnO_2@CRG is much less than that of SnO_2@TRG, as confirmed by XPS and XRF results (Table 6.1). In the post-graphenization process, due to π–π interaction between graphene sheets, it is likely to form agglomerations by restacked of reduced graphene oxide sheets during the solution-based reduction procedure. Meanwhile, the electrostatic repulsion originating from negatively charged functional groups, which keeps the graphene oxide hydrosol stable, is decreased [44, 45]. After the final drying process, the agglomeration becomes even more severe due to the capillary attraction effect, as water molecules are spilled out from the graphene interlayer [46]. However, the pre-introduced SnO_2 species on the basal plane of graphene oxide could act as the spacers between graphene so as to prevent the over-compact restacking and agglomeration of graphene during the wet process of reduction and drying.

6.3.3 *Electrochemcial Performance*

The SnO_2@TRG and SnO_2@CRG hybrids exhibit distinct microstructures (e.g., pore structure, particle loading state, composite interfacial property, and functionalities) for their different origin of fabrication processes (pre- or post-graphenizaton). It would be meaningful to correlate these structural differences with their electrochemical performance as energy storage materials. Thus, SnO_2@TRG and SnO_2@CRG, with the bare samples TRG and CRG as references, are fabricated and evaluated as electrodes for Li-ion batteries and supercapacitors, respectively.

6.3.3.1 Li-Ion Battery

SnO_2 shows a high theoretical capacity of Li^+ intercalations–deintercalations (790 mA h g^{-1}), which makes it a promising anode material for Li-ion battery. Figure 6.6 shows the representative CVs of the sample. Specifically, two pairs of redox current peaks can be clearly observed. The first dominant pair (cathodic, anodic) shown at the potential (V) of (0.01, 0.7) can be attributed to the alloying (cathodic scan) and dealloying (anodic scan) processes. The first pair is much more pronounced than the second pair, marking its major contribution to the total capacity of the cell. The intensity of this pair of SnO_2@TRG is much higher than that of SnO_2@CRG, which is attributed to higher loading of SnO_2 on TRG (2.17 at.%) than that on CRG (0.94 at.%). The second pair at (0.65, 1.3 V) is mainly appeared on the SnO_2@TRG electrode. This pair of redox peaks is related to the irreversible reduction of SnO_2 to Sn, which disappeared in the second cycle. Comparing with the first

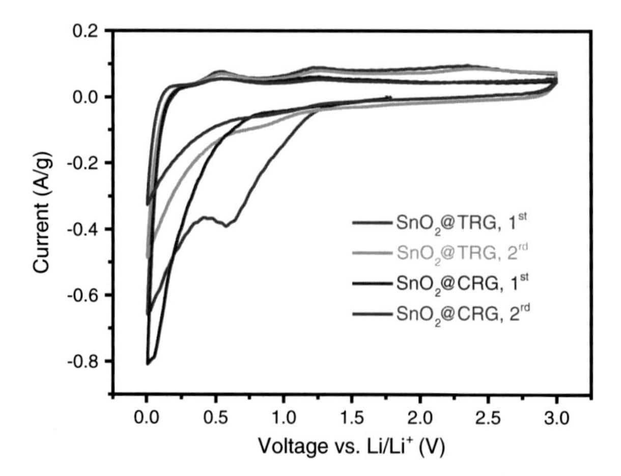

Fig. 6.6 The cyclic voltammograms of the SnO_2@T/CRG nanocomposite at a scan rate of 0.1 mV s^{-1}. Reproduced from Ref. [41] by permission of The Royal Society of Chemistry

cycle, the decreased intensity is mainly attributed to the reaction of oxygen-containing functional groups on graphene with lithium ions and the formation of surface polymeric layer due to the decomposition of the solvent in the electrolyte.

The rate performances of SnO_2@TRG and SnO_2@CRG are shown in Fig. 6.7. The discharge capacities of the first cycle are 1468 mA h g^{-1} for SnO_2@TRG, and 978 mA h g^{-1} for SnO_2@CRG. For the second cycle, the discharge capacities are 857 mA h g^{-1} for SnO_2@TRG, and 375 mA h g^{-1} for SnO_2@CRG. The formation of solid electrolyte interface (SEI) is the main reason that caused the large irreversible capacity loss. The discharge capacities in the sixth cycle are 707 mA h g^{-1} for SnO_2@TRG with a coulombic efficiency of 90 %, and 366 mA h g^{-1} for SnO_2@CRG with a coulombic efficiency of 94 %. With higher current density during the galvanostatic discharge (Li insertion, voltage decrease)/charge (Li extraction, voltage increase) process, the discharge capacities further decrease. The discharge capacities in the seventh cycle (current density at 400 mA g^{-1}) are 580 mA h g^{-1} for SnO_2@TRG, and 366 mA h g^{-1} for SnO_2@CRG. Further increasing of charge and discharge current to 800 mA g^{-1} caused the drop of capacity of SnO_2@TRG and SnO_2@CRG to 286 and 206 mA h g^{-1}, respectively. Upon returning back to a lower current density of 100 mA g^{-1}, the discharge capacities for SnO_2@TRG and SnO_2@CRG recover to 540 and 271 mA h g^{-1} in the 17th cycle.

As shown in Fig. 6.8, the SnO_2@graphene composite shows stable cycling performances. With a charge/discharge current density at 400 mA g^{-1}, SnO_2@TRG and SnO_2@CRG composites show discharge capacities of 204 and 196 mA h g^{-1} even after 100 cycles. A fast capacity loss is observed during the initial three cycles for SnO_2@CRG and CRG. Conversely, a gradual loss occurred on SnO_2@TRG and TRG. The higher initial discharge capacity of the SnO_2@TRG composite can be attributed to its loading amount of SnO_2 showed in Table 6.1. However, because of

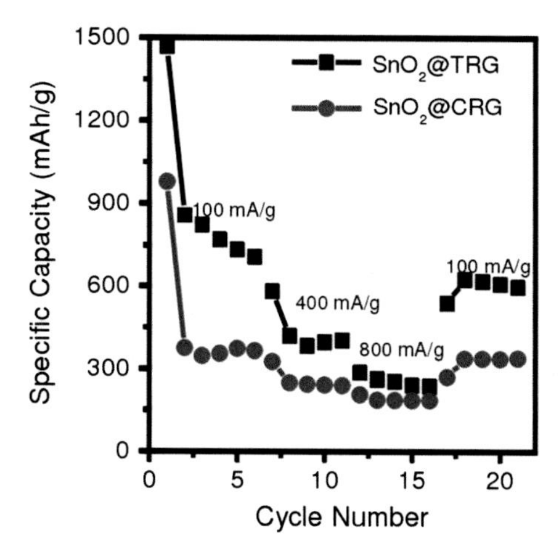

Fig. 6.7 The discharge capacity at different current rates for SnO_2@TRG and SnO_2@CRG nanocomposites. Reproduced from Ref. [41] by permission of The Royal Society of Chemistry

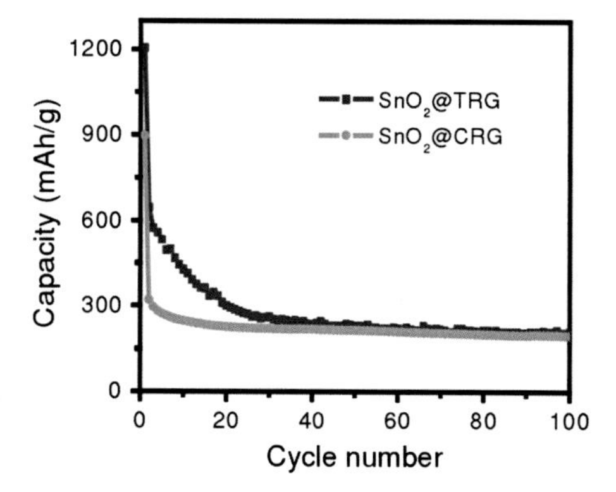

Fig. 6.8 The discharge capacity versus cycle number for two different samples at a charge-discharge current of 400 mA g^{-1}. Reproduced from Ref. [41] by permission of The Royal Society of Chemistry

the unstable porous structure and complex surface chemistry, the high capacity of both SnO_2@TRG and TRG is not well maintained. This indicated that pre-graphenization provides huge surface area for more metal oxide anchoring, while such a porous TRG is not very stable. In contrast, the post-graphenization process provides more anchoring sites for the formation of a large number of small SnO_2 particles, and the as-obtained SnO_2@CRG shows a more stable behavior as anode material.

6.3.3.2 Supercapacitor

Cyclic voltammetry (CV), electrochemical impedance spectrum (EIS), and galvanostatic charge/discharge (GC) techniques are employed to characterize the supercapacitor performance. As shown in Fig. 6.9a, at a very low scan rate of 3.0 mV s^{-1}, the CV curve of each sample indicates a typical electric double-layer capacitor (EDLC) character. Among them, SnO_2@CRG exhibits a more prominent Faradic redox reaction behavior with relatively lower resistance. Figure 6.9c shows the CV characteristics of SnO_2@CRG at different scanning rates, which maintain an EDLC behavior at the higher scan rates of over 200 mV s^{-1}. The gravimetric capacitance (C_F, F g^{-1}) of each electrode at various scan rates is calculated from CV curves and presented in Fig. 6.9b, while the initial C_F and associated specific capacitance (C_S, F m^2) at 3 mV s^{-1} is shown in Fig. 6.9d. It is found that the bare samples TRG and CRG exhibit quite a similar capacitance (162.1 and 169.3 F g^{-1}). However, as the scan rate increases, the value of CRG drops very fast with a final retention of 22.6 % at 500 mV s^{-1}, which is significantly lower than that of TRG (35.1 % at 500 mV s^{-1}). This phenomenon is in accordance with the microstructure and pore structure of thermally/chemically reduced graphene. On one hand, with the exterior porous structure of the assembly and residual surface oxygen functionalities on each basic building block, TRG is endowed with the active surface areas which are highly accessible to the electrolyte. Conversely, the inner surface of ink-bottle pores in CRG, which greatly contribute to the BET surface area, is very difficult to be wetted by the electrolyte. Therefore, though the BET area of TRG (293 m^2 g^{-1}) is significantly smaller than that of CRG (666 m^2 g^{-1}), the specific capacitance per BET area (C_S) of TRG (0.55 F m^{-2}) is remarkably higher than that of CRG (0.25 F m^{-2}). On the other hand, the 'open' structure of TRG with numerous large pores could serve as the buffer pool to the electrolyte, thus providing a convenient path for ion diffusion between the graphene sheets and electrolyte. Therefore, the TRG electrode exhibits a relatively quicker response in charge/discharge cycling than CRG constructed by over compacted graphene sheets.

The TRG and CRG behave quite differently as supercapacitor electrodes after introduction of SnO_2 nanoparticles. As shown in Fig. 6.9d, both the initial C_F at 3 mV s^{-1} and the capacitance retention at higher scan rates of SnO_2@CRG are higher than that of SnO_2@TRG. However, the supercapacitor performance of the SnO_2@TRG electrode decreased compared with the bare TRG. As a result, the SnO_2@CRG exhibits the highest C_F (189.4 F g^{-1} at 3.0 mV s^{-1}) as a supercapacitor electrode among all the samples, with the retention of 33.7 %. The controversial effect of SnO_2 hybridization on supercapacitor performance is primarily dependent on the pre- or post-graphenization strategy induced structural difference of the resulting materials. The GC curves at the current density of 1.0 A g^{-1} are shown in Fig. 6.10a. All samples exhibit typical symmetrical charge-discharge patterns. In accordance with the results calculated from CV curves, SnO_2@CRG owns the best performance (184.0 F g^{-1}) on energy storage among the four samples.

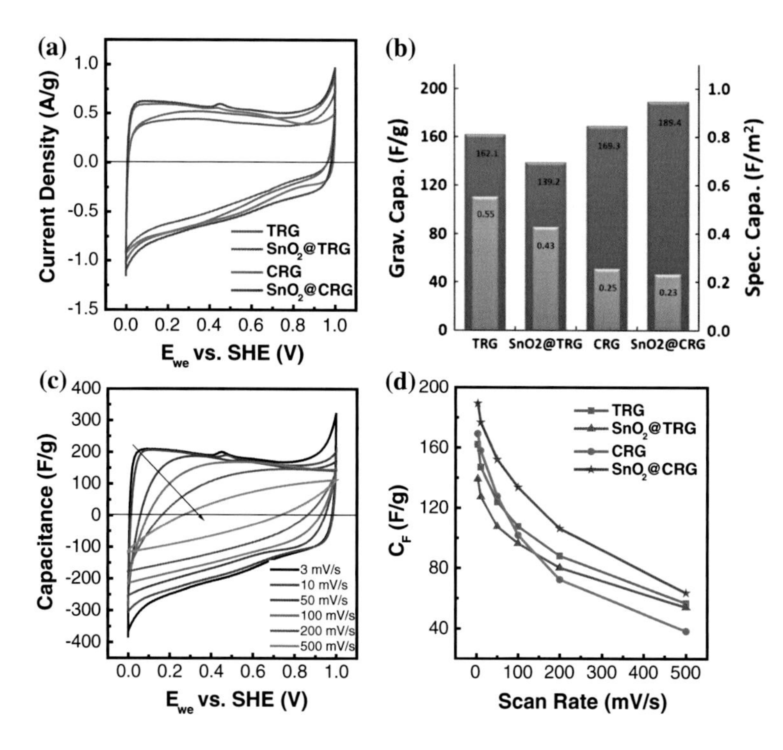

Fig. 6.9 **a** CV curves and **b** as-calculated CF and CS of TRG, SnO_2@TRG, CRG and SnO_2@CRG at 3 mV s^{-1}; **c** CV evolution of SnO_2@CRG with scan rates varies from 3.0 to 500 mV s^{-1}; **d** CF of four samples at different scan rates from 3.0 to 500 mV s^{-1}. Reproduced from Ref. [41] by permission of The Royal Society of Chemistry

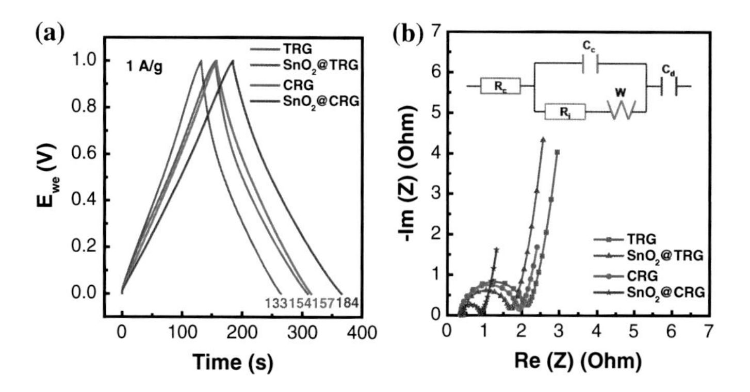

Fig. 6.10 **a** GC spectrum and **b** Nyquist plots of EIS for TRG, SnO_2@TRG, CRG, and SnO_2@CRG. Reproduced from Ref. [41] by permission of The Royal Society of Chemistry

Table 6.2 Summary of internal components in electrodes

Sample	R_i (Ω)	R_c (Ω)	C_c (F/g)	W (Ω $S^{-1/2}$)	C_d (F/g)
TRG	1.65	0.40	0.27	0.71	170.51
SnO_2@TRG	1.22	0.41	0.15	0.74	145.72
CRG	1.50	0.36	0.12	0.45	226.07
SnO_2@CRG	0.51	0.36	0.09	0.36	232.53

Reproduced from Ref. [41] by permission of The Royal Society of Chemistry

To further confirm the double-layer formation of the electrodes, AC impedance spectroscopy is employed to determine the internal components of the devices (Nyquist plots as shown in Fig. 6.10b). An equivalent circuit model (inset of Fig. 6.10b) is introduced to simulate the capacitive and resistive elements of the cells under analysis. These elements include the internal resistance of the electrode (R_i), the capacitance and resistance of the contact interface (C_c and R_c), a Warburg diffusion element attributable to the ion migration through the electrode (Z_w), and the capacitance inside the pores (C_d) [47, 48]. The fitting results are shown in Table 6.2. In accordance with the synthesis strategy-induced structural difference, the as-obtained SnO_2@graphene hybrids by either pre-graphenization or post-graphenization exhibit a decrease in R_i, due to the increase of charge carrier density in graphene lattice from the electron donor-SnO_2 nanoparticles. As a result, SnO_2@CRG has the minimum R_i (0.51 Ω) among all the samples. However, the ion diffusion behavior shows distinct tendency upon SnO_2 decoration by two different graphenization strategies. Compared with TRG (0.71 Ω $s^{1/2}$), the Warburg diffusion resistance of SnO_2@TRG is increased to 0.74 Ω $s^{1/2}$, which is ascribed to the collapse of some honeycomb-like structures after impregnation and drying process. Conversely, the value for SnO_2@CRG (0.36 Ω $s^{1/2}$) is smaller than that of CRG (0.45 Ω $s^{1/2}$), which is attributed to the prominent 'spacer' effect of SnO_2 nanoparticles. Thereafter, from a view-point of thermodynamics, the mass transfer within the electrochemical system is significantly promoted through: (a) improvement of charge transfer along the graphene lattice and between graphene by π–π interactions of delocalized electrons arising from both graphitic domains and SnO_2 electron donors; and (b) optimization of the porous structure by preventing the restacking of graphene so as to offer a low resistance channel for ion diffusion. Thus, the PC-active SnO_2 species combining with the graphene surface inside the pores of SnO_2@CRG are highly accessible to the ions of electrolyte, so as to give rise to capacitance inside pores with a maximum value of as high as 232.53 F g^{-1} among the four samples.

6.4 Conclusions

To clarify the effect of pre- or post-graphenization for graphene-based hybrids, two kinds of SnO_2@graphene hybrids are explored with the same GO and Sn precursors. The SnO_2 nanoparticles with a loading of 6.3 wt% are distributed on porous TRG, while a loading of 3.3 wt% was determined on compacted CRG. When served as supercapacitor electrode, SnO_2@CRG exhibits the highest C_F among all the samples with an initial C_F approaching 189.4 F g^{-1} and retention of 33.7 % at a scan rate of 3.0 mV s^{-1}. The gravimetric capacitance of CRG drops rapidly with a final retention of 22.6 % at 500 mV s^{-1}, which is significantly lower compared with TRG (35.1 % at 500 mV s^{-1}). When they are employed as Li-ion battery electrode, the discharge capacities in the first cycle are 1468 and 978 mA h g^{-1} for SnO_2@TRG and SnO_2@CRG, respectively. During the galvanostatic discharge (Li insertion, voltage decreases)/charge (Li extraction, voltage increases) process, the discharge capacities of both SnO_2@TRG and SnO_2@CRG decreased gradually with the increase of current density. Even after 100 cycles at 400 mA g^{-1}, a discharging capacity of 204 and 196 mA h g^{-1} can still be retained for SnO_2@TRG and SnO_2@CRG, respectively. This suggested that the method of graphenization provides the hybrid with different structure and electrochemical performance. The pre-graphenization process provides a large amount of pores for ion diffusion for supercapacitors and Li-ion batteries, but poor stability, while the post-graphenization process offers compact graphene and good interaction between the SnO_2 and graphene, which is stable during long-term cycling of supercapacitor and Li-ion battery. The optimized graphenization method for graphene hybrids should be further explored to provide new insights into hybrid formation and advanced materials for energy storage.

References

1. Yan J, Fan ZJ, Wei T, Qian WZ, Zhang ML, Wei F. Fast and reversible surface redox reaction of graphene-MnO_2 composites as supercapacitor electrodes. Carbon. 2010;48(13):3825–33.
2. Frackowiak E, Beguin F. Carbon materials for the electrochemical storage of energy in capacitors. Carbon. 2001;39(6):937–50.
3. Kaskhedikar NA, Maier J. Lithium storage ion carbon nanostructures. Adv Mater. 2009;21(25–26):2664–80.
4. Arico AS, Bruce P, Scrosati B, Tarascon JM, Van Schalkwijk W. Nanostructured materials for advanced energy conversion and storage devices. Nat Mater. 2005;4(5):366–77.
5. Bruce PG, Scrosati B, Tarascon JM. Nanomaterials for rechargeable lithium batteries. Angew Chem Int Ed. 2008;47(16):2930–46.
6. Liu C, Li F, Ma LP, Cheng HM. Advanced materials for energy storage. Adv Mater. 2010;22(8):E28–62.
7. Su DS, Schlogl R. Nanostructured carbon and carbon nanocomposites for electrochemical energy storage applications. ChemSusChem. 2010;3(2):136–68.
8. Huang C-H, Doong R-A. Sugarcane bagasse as the scaffold for mass production of hierarchically porous carbon monoliths by surface self-assembly. Micropor Mesopor Mat. 2012;147(1):47–52.
9. Huang CH, Doong RA, Gu D, Zhao DY. Dual-template synthesis of magnetically-separable hierarchically-ordered porous carbons by catalytic graphitization. Carbon. 2011;49(9):3055–64.
10. Zhang Q, Huang JQ, Zhao MQ, Qian WZ, Wei F. Carbon nanotube mass production: Principles and processes. ChemSusChem. 2011;4(7):864–89.
11. Zhu YW, Murali S, Cai WW, Li XS, Suk JW, Potts JR, et al. Graphene and graphene oxide: synthesis, properties, and applications. Adv Mater. 2010;22(35):3906–24.

12. Sun YQ, Wu QO, Shi GQ. Graphene based new energy materials. Ener Environ Sci. 2011;4(4): 1113–32.
13. Guo SJ, Dong SJ. Graphene nanosheet: synthesis, molecular engineering, thin film, hybrids, and energy and analytical applications. Chem Soc Rev. 2011;40(5):2644–72.
14. Bai H, Li C, Shi GQ. Functional composite materials based on chemically converted graphene. Adv Mater. 2011;23(9):1089–115.
15. Rao CNR, Sood AK, Subrahmanyam KS, Govindaraj A. Graphene: the new two-dimensional nanomaterial. Angew Chem Int Ed. 2009;48(42):7752–77.
16. Hu YH, Wang H, Hu B. Thinnest two-dimensional nanomaterial-graphene for solar energy. ChemSusChem. 2010;3(7):782–96.
17. Liang MH, Zhi LJ. Graphene-based electrode materials for rechargeable lithium batteries. J Mater Chem. 2009;19(33):5871–8.
18. Zhu Y, Murali S, Cai W, Li X, Suk JW, Potts JR, et al. Graphene and graphene oxide: Synthesis, properties, and applications. Adv Mater. 2010;22(35):3906–24.
19. Cheng Q, Tang J, Ma J, Zhang H, Shinya N, Qin LC. Graphene and nanostructured MnO_2 composite electrodes for supercapacitors. Carbon. 2011;49(9):2917–25.
20. Fan ZJ, Yan J, Wei T, Zhi LJ, Ning GQ, Li TY, et al. Asymmetric supercapacitors based on graphene/MnO_2 and activated carbon nanofiber electrodes with high power and energy density. Adv Funct Mater. 2011;21(12):2366–75.
21. Yang SB, Feng XL, Ivanovici S, Mullen K. Fabrication of graphene-encapsulated oxide nanoparticles: Towards high-performance anode materials for lithium storage. Angew Chem Int Ed. 2010;49(45):8408–11.
22. Yang SB, Cui GL, Pang SP, Cao Q, Kolb U, Feng XL, et al. Fabrication of cobalt and cobalt oxide/graphene composites: towards high-performance anode materials for lithium ion batteries. ChemSusChem. 2010;3(2):236–9.
23. Kim H, Seo DH, Kim SW, Kim J, Kang K. Highly reversible Co_3O_4/graphene hybrid anode for lithium rechargeable batteries. Carbon. 2010;49(1):326–32.
24. Liu H, Wang GX, Liu J, Qiao SZ, Ahn HJ. Highly ordered mesoporous NiO anode material for lithium ion batteries with an excellent electrochemical performance. J Mater Chem. 2010;21(9): 3046–52.
25. Yao J, Shen XP, Wang B, Liu HK, Wang GX. In situ chemical synthesis of SnO_2-graphene nanocomposite as anode materials for lithium-ion batteries. Electrochem Commun. 2009;11(10):1849–52.
26. Kim H, Kim SW, Park YU, Gwon H, Seo DH, Kim Y, et al. SnO_2/graphene composite with high lithium storage capability for lithium rechargeable batteries. Nano Res. 2010;3(11):813–21.
27. Li FH, Song JF, Yang HF, Gan SY, Zhang QX, Han DX, et al. One-step synthesis of graphene/ SnO_2 nanocomposites and its application in electrochemical supercapacitors. Nanotechnol. 2009;20(45):455602.
28. Paek SM, Yoo E, Honma I. Enhanced cyclic performance and lithium storage capacity of SnO_2/graphene nanoporous electrodes with three-dimensionally delaminated flexible structure. Nano Lett. 2009;9(1):72–5.
29. Lian PC, Zhu XF, Liang SZ, Li Z, Yang WS, Wang HH. High reversible capacity of SnO_2/ graphene nanocomposite as an anode material for lithium-ion batteries. Electrochim Acta. 2011;56(12):4532–9.
30. Song HJ, Zhang LC, He CL, Qu Y, Tian YF, Lv Y. Graphene sheets decorated with SnO_2 nanoparticles: in situ synthesis and highly efficient materials for cataluminescence gas sensors. J Mater Chem. 2011;21(16):5972–7.
31. Zhang M, Lei D, Du ZF, Yin XM, Chen LB, Li QH, et al. Fast synthesis of SnO_2/graphene graphene oxide with stannous ions. J Mater Chem. 2011;21(6):1673–6.
32. Li YM, Lv XJ, Lu J, Li JH. Preparation of SnO_2-nanocrystal/graphene-nanosheets composites and their lithium storage ability. J Phys Chem C. 2010;114(49):21770–4.

33. Zhang LS, Jiang LY, Yan HJ, Wang WD, Wang W, Song WG, et al. Mono dispersed SnO_2 nanoparticles on both sides of single layer graphene sheets as anode materials in Li-ion batteries. J Mater Chem. 2010;20(26):5462–7.
34. Wang XY, Zhou XF, Yao K, Zhang JG, Liu ZP. A SnO_2/graphene composite as a high stability electrode for lithium ion batteries. Carbon. 2011;49(1):133–9.
35. Huang XD, Zhou XF, Zhou LA, Qian K, Wang YH, Liu ZP, et al. A facile one-step solvothermal synthesis of SnO_2/graphene nanocomposite and its application as an anode material for lithium-ion batteries. ChemPhysChem. 2011;12(2):278–81.
36. Wang JZ, Zhong C, Wexler D, Idris NH, Wang ZX, Chen LQ, et al. Graphene-encapsulated Fe_3O_4 nanoparticles with 3D laminated structure as superior anode in lithium ion batteries. Chem Eur J. 2011;17(2):661–7.
37. Zhao XC, Wang AQ, Yan JW, Sun GQ, Sun LX, Zhang T. Synthesis and electrochemical performance of heteroatom-incorporated ordered mesoporous carbons. Chem Mater. 2010;22(19):5463–73.
38. Zhang K, Zhang LL, Zhao XS, Wu JS. Graphene/polyaniline nanoriber composites as supercapacitor electrodes. Chem Mater. 2010;22(4):1392–401.
39. Liu AR, Li C, Bai H, Shi GQ. Electrochemical deposition of polypyrrole/sulfonated graphene composite films. J Phys Chem C. 2010;114(51):22783–9.
40. Fan ZJ, Yan J, Zhi LJ, Zhang Q, Wei T, Feng J, et al. A three-dimensional carbon nanotube/graphene sandwich and its application as electrode in supercapacitors. Adv Mater. 2010;22(33):3723–8.
41. Chen CM, Zhang Q, Huang JQ, Zhang W, Zhao XC, Huang CH, et al. Chemically derived graphene-metal oxide hybirds as electrodes for electrochemical energy storage: pre-graphenization or post-graphenization? J Mater Chem. 2012;22:13947–55.
42. Lv W, Tang DM, He YB, You CH, Shi ZQ, Chen XC, et al. Low-temperature exfoliated graphenes: Vacuum-promoted exfoliation and electrochemical energy storage. ACS Nano. 2009;3(11):3730–6.
43. Inagaki M. Pores in carbon materials-Importance of their control. New Carbon Mater. 2009;24(3):193–222.
44. Cote LJ, Kim F, Huang J. Langmuir-Blodgett assembly of graphite oxide single layers. J Am Chem Soc. 2009;131(3):1043–9.
45. Kim J, Cote LJ, Kim F, Yuan W, Shull KR, Huang JX. Graphene oxide sheets at interfaces. J Am Chem Soc. 2010;132(23):8180–6.
46. Yang X, Zhu J, Qiu L, Li D. Bioinspired effective prevention of restacking in multilayered graphene films: Towards the next generation of high-performance supercapacitors. Adv Mater. 2011;23(25):2833–8.
47. Xu GH, Zheng C, Zhang Q, Huang JQ, Zhao MQ, Nie JQ, et al. Binder-free activated carbon/carbon nanotube paper electrodes for use in supercapacitors. Nano Research. 2011;4(9):870–81.
48. Huang CW, Hsu CH, Kuo PL, Hsieh CT, Teng HS. Mesoporous carbon spheres grafted with carbon nanofibers for high-rate electric double layer capacitors. Carbon. 2011;49(3):895–903.

Chapter 7
Main Conclusions and Plan of Further Work

7.1 Conclusions

1. Thermally reduced graphene (TRG) with tailored surface chemistry was prepared by vacuum promoted thermal expansion of GO, followed by annealing at different temperatures. In one aspect, the C=O components (mainly contributed by carboxyls and carbonyls) are dramatically decreased from 8.71 at.% (GO) to only 1.40 at.% (G250), while the C(O)O related components (mainly ascribed to anhydrides and lactones) are simultaneously increased from 0.66 at.% (GO) to 1.52 at.% (G250) during annealing at 250 °C. After carbonization at 1000 °C, most oxygen components in TRG have been removed. The O–H, C–O, and C=O related functional groups, which are respectively attributed to thermally more stable isolated phenols, ethers, and carbonyls, and finally become the major components in G1000. In another aspect, the desorption of oxygen-bonded carbons as CO_2 and CO generates large amount of topological defects and vacancies in the graphene lattice, with a simultaneous 'self-healing' of graphitic lattice ('re-graphitization'). The TRGs were further fabricated into and characterized as supercapacitor electrodes; it is found that the thermal annealing-induced structural evolution play a key role in determining the electrochemical performance of TRGs toward supercapacitor applications. The oxygen functional groups can significantly enhance the capacitance performance of TRGs by introducing abundant PC-active sites through reversible Faradic redox reactions. These results, which correlate the surface chemistry and structural properties of TRGs to the supercapacitor performances, provide a reliable experimental and theoretical model for the development of graphene-based electrode materials with controllable properties, and will benefit the application of graphene in advanced energy storage.
2. Hierarchically, aminated graphene honeycombs are obtained through vacuum assisted thermal expansion of graphite oxide followed by amination. Working as energy storage material in electrochemical capacitor, the large specific area

C.-M. Chen, *Surface Chemistry and Macroscopic Assembly of Graphene for Application in Energy Storage*, Springer Theses,
DOI 10.1007/978-3-662-48676-4_7

contributes for electrical double layer capacitance, while the rational 3D macroporous structure provides cushion space for electrolyte and acts as low-resistance channel for ion transfer. When the amination temperature is low (such as 200 °C), NH_3 reacts with carboxylic acid species to form mainly intermediate amide or amine like species ('chemical N') through nucleophilic substitution. When the amination temperature increases, the intramolecular dehydration or decarbonylation will take place to generate thermally stable heterocyclic aromatic moieties, such as pyridine, pyrrole, and quaternary type N sites ('lattice N'). This functional 3D assembly of graphene shows a promising prospect in electrochemical capacitive energy storage with high capacitance (e.g., gravimetric capacitance 207 F g^{-1} and specific capacitance 0.84 F m^{-2} at 3 mV s^{-1}) and long cycle-life (e.g., 97.8 % of retention rate after 3000 cycles, and maintains 47.8 % of capacitance at 500 mV s^{-1} comparing with 3 mV s^{-1}). Such porous hierarchical architectures will benefit applications in heterogeneous catalysis, separation, and drug delivery, which requires fast mass transfer through mesopores, reactant reservoirs, and tunable surface chemistry.

3. A highly conductive, free-standing graphene film was obtained by annealing GO film between two stacked wafers. The oxygen-containing functional groups were removed during annealing, which caused the decrease of the sp^2 domains size. The conjugation of the graphene basal plane and π–π interactions between GO and reduced GO sheets favor the transport of charge carriers, and render the free-standing RGF with good electric conductivity. This is an efficient and effective way to fabricate highly conductive RGF in large scale and low cost, which provides graphene macroscopic materials for application exploration.
4. A 3D macroporous bubble graphene film, with hollow graphene bubble serving as the replicated building block, was obtained by a facile hard template-directed ordered assembly approach. The graphene oxide precursor and template PMMA spheres were orderly assembled by vacuum filtration, followed by the calcination to reduce graphene oxide into graphene, as well as to remove the template via pyrolysis. The as-obtained free-standing film afforded quite high rate capability (1.0 V s^{-1}) as a 'binder-free' supercapacitor electrode. This work provides a simple and green synthetic strategy to obtain free-standing macroporous graphene assemblies with tunable microstructure, which shows promising prospects toward other applications in biology, energy conversion, and catalysis.
5. To clarify the effect of pre- or post-graphenization for graphene-based hybrids, two kinds of SnO_2@graphene hybrids are explored with the same GO and Sn precursors. The SnO_2 nanoparticles with a loading of 6.3 wt% are distributed on porous TRG, while a loading of 3.3 wt% was determined on compacted CRG. When the SnO_2@graphene is used as a supercapacitor electrode, the SnO_2@CRG exhibits the highest C_F as a supercapacitor electrode among all the samples with an initial C_F approaching 189.4 F g^{-1} and retention of 33.7 % at 3.0 mV s^{-1}. The gravimetric capacitance of CRG drops very fast with a final retention of 22.6 % at 500 mV s^{-1}, which is significantly lower compared with TRG (35.1 % at 500 mV s^{-1}). When they are employed as Li-ion battery electrodes, the discharge capacities in the first cycle are 1468 and 978 mA h g^{-1} for SnO_2@TRG and

SnO_2@CRG, respectively. With higher current density during the galvanostatic discharge (Li insertion, voltage decreases)/charge (Li extraction, voltage increases) process, the discharge capacities of both SnO_2@TRG and SnO_2@CRG decreased gradually. Even after 100 cycles at 400 mA g^{-1}, a discharging capacity of 204 and 196 mA h g^{-1} can still be retained for SnO_2@TRG and SnO_2@CRG, respectively. This suggested that the method of graphenization provides the hybrid with different structure and electrochemical performance. The pre-graphenization process provides a large amount of pores for ion diffusion, which is of benefit for loading of SnO_2, fast ion diffusion for supercapacitors, and higher capacity for Li-ion batteries, but poor stability, while the post-graphenization process offers compact graphene and good interaction between the SnO_2 and graphene, which provides stable structure for long term stability for supercapacitor and Li-ion battery use.

7.2 Primary Innovation Points

1. Bridge the gap between graphene's surface chemistry and electrochemical performances
   The microstructure and surface chemistry evolving mechanism, during annealing and aminating of TRG at different temperatures, were studied systematically and deeply for the first time, together with correlating with the electrochemical performances of these materials. It laid theoretical foundation for controllable large-scale production and surface modification of reduced graphene, and provided new understanding in design and development of electrode material for graphene-based supercapacitor. Different composite ways were also researched for the effects on the microstructure of graphene supported nanoparticles, establishing the relationship with application performances of supercapacitor and Li-ion batteries. The work provided solid theoretical basis for developing graphene-based energy storage devices with diversified functionalities.
2. Design and assembly of graphene architectures
   For the applications in high electrical and thermal conductivity, an original self-assembly process at the liquid/air interface was used to prepare free-standing graphene oxide film, combining with annealing in a confined space between two stacked substrates to form a free-standing highly conductive graphene film. Graphene sheets were overlapped orderly in this process. For the applications in flexible energy storage with high power density, template-directed assembly approach was used to obtain 3D macroporous graphene film with controllable pore size and microstructure. The work provided a universal method for developing new nanostructured graphene material with the integration of structure/functionality, which paved the way for commercial application of graphene in new energy and electronics.

7.3 Planning of Future Work

Similar to single wall carbon nanotube and fullerene, graphene is a new member of carbon materials. Graphene exhibits many excellent physical chemistry performances, including mechanical, thermal, light, and electrical, and high theoretical specific surface area. However, the typical sp^2 conjugated structure gives it some unusual electronic and quantum characters. In addition, different from the preparation of traditional nano carbon materials with costly chemical vapor deposition or arc discharge method, graphene can be prepared by chemical intercalation, exfoliation, and reduction. This preparation method uses inexpensive graphite as raw material, realizing large-scale production with low cost. The chemically derived graphene possesses high-specific surface area, and inherits controllable surface chemistry, providing facility for further modification, assembly or composite. Currently, chemically derived graphene has become superior nano material with great developing potential and industrialization advantage, showing promising prospect in microelectronic, energy storage and conversion, and heterocatalysis, etc. The following aspects are estimated to be the research focus of graphene in next 5 years, which should be paid special attention.

(1) Technical problems of chemically prepared graphene in large scale: green development of preparing graphite oxide in low cost and high efficiency, process optimization, and control of chemical conversion from graphite oxide to graphene, improvement of thermal expansion process to obtain chemically derived graphene samples with high quality.
(2) Design and development of graphene architecture: assembly of high thermal conductive free-standing graphene film in large area for the application in thermal management, development of transparent, and conductive ultrathin window materials for the applications in solar battery and touch screen, graphene foam-based energy storage and thermal insulation materials with controllable pore structure, high-specific surface area, and high activity.
(3) Research and development of graphene catalytic chemistry: development and evaluation of new catalyst supporting system to replace the traditional carbon materials, electrochemical catalysis of graphene in oxygen reduction, methyl alcohol and formic acid electrooxidation, water oxidation, electrochemical synthesis etc.

Both the research depth and breadth should be noted. On the one hand, the substantive law behind experimental phenomenon needs to be explored deeply, providing right theoretical foundation for controlling material performances. On the other hand, it needs to broaden one's mind and introduce new thought to get related experimental methods and research advance in nano field, and then apply and promote these to developing graphene, broadening its application prospects.

Printed in the United States
By Bookmasters